Basic Quantum Chemistry

Basic Quantum Chemistry

Leon F. Phillips

Department of Chemistry, University of Canterbury

John Wiley & Sons, Inc., New York · London · Sydney

For Pam and John

Preface

The main intention of this book is to provide the theoretical knowledge which it is becoming more and more necessary for all but the most classically oriented chemist to possess. The treatment is tailored particularly to the needs of the student or research worker on the experimental side of chemistry, but it might also serve as a useful, if nonrigorous, introduction to the field for a budding theoretician. In either case the aim is to carry the reader from the sort of qualitative understanding that is commonly attained in undergraduate courses on valence theory to the stage where he is able to appreciate the importance of new theoretical advances, to read much of the current literature of theoretical chemistry, and to have access to more advanced and specialized texts, some of which are listed in Appendix I. The descriptive word "basic" in the title is therefore reasonably accurate, not in the sense that no previous acquaintance with quantum theory is assumed, but in the sense that the book contains what is probably to be regarded as essential at the present time.

The contents of the book are based on a series of lectures which have been devised for fourth year chemistry students at the University of Canterbury. These students concurrently take advanced lectures in physical, organic, and inorganic chemistry, and in the relatively descriptive general courses they make considerable use of the theoretical results, and also of the possibility of reinforcing new concepts by viewing them in relation to several different sets of facts in the various courses. By the time they have reached fourth year level the students are selected more for tenacity than on the basis of mathematical ability, and it is hoped that the same may be true of the readers of this book.

The structure of the book corresponds quite closely to that of the lecture course in that the first part (up to the end of Chapter 4) contains the theory which is required in order to handle the applications that are discussed in the last three chapters. These applications fall, conveniently enough, into each of the three main branches of chemistry. The major topics are, in order, Hückel molecular orbital theory (organic), ligand field theory (inorganic), and some aspects of spectroscopy (physical). A feature of these applications is the use which is customarily made of deductions from symmetry properties, and the necessary group theory is developed from first principles in Chapter 4. A few other mathematical points are worth mentioning, the first being that use is frequently made of matrix notation, generally in order to simplify the writing down of linear equations. A detailed knowledge of matrix algebra is not required in order to employ the notation in this way, and in the few places where results that are actually derived from matrix algebra are needed they are properly signposted. Very similar remarks can be made in connection with the use of operators, eigenfunctions, and eigenvalues, all of which sound much more complicated than they are. A final point is that certain mathematical derivations, notably those in Chapters 2 and 3, are regarded as being of the nature of background material to which the student should be exposed more or less lightly. This is in contrast to the material which is discussed in the last three chapters, where the essential thing is to be able to use the methods that are outlined and apply them to new problems.

It is a pleasure to acknowledge the assistance toward the writing of this book which I have received from many of my colleagues at the University of Canterbury, especially Dr. M. T. Christensen, who is to be held largely responsible for any clarity that has crept into the first three chapters, Mr. J. C. W. De la Bere who provided helpful criticism of Chapter 4, and Dr. A. Fischer and Dr. J. E. Fergusson who have similarly criticized Chapters 5 and 6.

Christchurch, New Zealand   Leon F. Phillips
February, 1965

Contents

1

Formal Quantum Theory

....

1a INTRODUCTION

This chapter is intended to form the theoretical foundation for the
remainder of the book. Here we shall define our vocabulary, both
verbal and mathematical, and try to obtain some feeling for the
kind of information which quantum theory can supply about a
physical or chemical system. Our approach will be to take for
granted the experimental evidence which is provided by spectroscopy or electron diffraction and to proceed by way of logical development from a minimum number of fundamental postulates. The
applications of the theory will be deferred until later, and it will be
assumed that the reader is already acquainted with its historical
development, from the discoveries of Planck, Einstein, and Bohr to
those of Heisenberg, de Broglie, Schroedinger, Pauli, Dirac, and
others.

In its modern form quantum mechanics resembles a system of
geometry in which certain axioms or postulates are stated and a
set of rules is provided to enable the consequences of the postulates
to be worked out. Because quantum mechanics is a part of physics
the postulates must be chosen in such a way that their consequences
are consistent with experimental observations, and much of the
interest and fascination of the theory derives from the fact that the
resulting postulates are of a very different nature from those which
form the basis of classical mechanics. It is, perhaps, surprising that
these postulates nevertheless give rise to a theory which contains
classical mechanics as a limiting case.

From our point of view the most serious assumption of classical mechanics is that any dynamic system can be regarded as an assemblage of particles whose masses, positions, and velocities can be determined precisely at any instant. From the laws of classical mechanics it then follows that it is possible (in principle) to deduce what the values of these physical quantities must be at any other instant, assuming only that the forces between the particles are known and that the equations of motion can be solved. In contrast with this clear-cut view of nature, the picture which is provided by quantum theory is remarkable for its blurred character. Thus it is both a characteristic of the theory and an experimental fact that we cannot decide with certainty the value of a physical quantity, such as the position of a particle at a particular instant, but in general we can only specify the probability distribution which would be obtained from a large number of measurements of the quantity in question. It is often possible to make the range of probable values of a particular quantity extremely narrow, but only at the expense of the precision with which some other variable of the system may be determined. (This is the situation which is described by the well-known Heisenberg uncertainty principle.)

In the limiting case, where a system begins to conform to the laws of classical mechanics, this uncertainty as to the precise values of physical quantities still remains, but for all practical purposes it is negligible. We can say that the classical limit is approached when the *action* of a system (product of energy and time) is very much greater than Planck's constant (6.624×10^{-27} erg. sec.). An equivalent statement is that a system behaves classically when the numbers of quanta involved (the "quantum numbers") are very large. We shall not discuss the transition between classical and quantum mechanics in any detail, but it is useful to note that the correspondence between the two cases is implied in the second of the fundamental postulates which are given in section 1b. We shall make use of the result of considering such a transition in connection with the discussion of the Hamiltonian operator in section 1d. In this connection it is interesting to consider that the loss, on going from the classical to the quantum mechanical description of a system, of the precise values of physical quantities is compensated for by the appearance of several phenomena which have no classical analogue, for example, electron spin and the tunnel effect.

An incidental feature of formal quantum theory, but one which can be very important from the student's point of view, is the extensive use which is made of linear operators with their concomitant eigenfunctions and eigenvalues. This usage can lead to some rather unfamiliar-looking mathematics, so that the first glance at a page of undiluted quantum mechanics may be sufficient to discourage a student for life. Fortunately, it is easy to overcome this unfamiliarity by working through some exercises, such as those at the end of this chapter, and it is then found that the algebraic manipulation of these quantities is quite straightforward and follows simple rules.

1b THE POSTULATES OF QUANTUM THEORY

We shall require only three fundamental postulates from which to develop the theory. These are as follows:

POSTULATE 1. A quantum mechanical system is described as completely as possible when a function Ψ, usually called the *wave function*, is specified. The wave function Ψ is required to be a finite, single-valued, and continuous function of the internal and external coordinates of the system. The description of the system takes the following form: If the coordinates are $q_1, q_2, \cdots q_n$, then the probability that they will have values which lie in the ranges q_1 to $q_1 + dq_1, q_2$ to $q_2 + dq_2 \cdots q_n$ to $q_n + dq_n$ respectively is equal to $\Psi^*\Psi \, d\tau$, where Ψ^* is the complex conjugate of Ψ and

$$d\tau = dq_1 \cdot dq_2 \cdots dq_n \tag{1.1}$$

is an element of volume in the configuration space of the system. A further restriction on the mathematical form of the wave function is that it should possess an integrable square, i.e., it must be possible to integrate $\Psi^*\Psi \, d\tau$ over the whole range of the coordinates q.

We observe that the wave function is in general complex, i.e., it contains the square root of minus one, but this need not disturb us because we shall find that the physical properties of systems all depend on the *real* quantity $\Psi^*\Psi$. We also note that the idea of probability is introduced in the first and most fundamental postulate of the theory. The consequences of this are correspondingly far reaching.

According to this first postulate the probability of finding a system in a given state depends on the square of the absolute value of the quantity Ψ for that state. This leads to the possibility of observing what are usually termed *interference effects*, in which the overlap of two regions where $|\Psi|^2$ is finite gives rise to a region in which $|\Psi|^2$ is zero, because the wave functions which describe the two regions happen to be of opposite sign. This is a typical wave phenomenon, and in fact the analogy of Ψ to the amplitude of a wave in classical mechanics is complete, because the differential equations which can be used to describe the two entities (a wave amplitude and a quantum mechanical wave function) are of identical form. (Cf. postulate 3.)

The choice of coordinates with which to characterize the system is somewhat arbitrary. In the main we shall use either cartesian coordinates (x, y, z, and time t) or spherical polar coordinates (r, θ, ϕ, and t).

POSTULATE 2. To every observable physical quantity of a system f, for example, there corresponds a linear operator[1] $\hat{f}$ such that

$$\hat{f}\Psi = f\Psi \qquad (1.2)$$

This is usually expressed in words by saying that f is an *eigenvalue* and Ψ is an *eigenfunction* of the operator $\hat{f}$. Equation (1.2) indicates that the physical quantity in question has the definite value f in the state described by Ψ.

This relationship enables us to extract from the wave function all the information which it contains as to the values of physical quantities in a system. Of course, to be able to do this we need to know both the form of the wave function for any given system and the nature of the linear operator which corresponds to a particular physical quantity. In section 1d we shall consider in principle how the form of the wave function may be determined; the practice of doing this, as distinct from the principle, constitutes a major part of

---

[1] A linear operator obeys the relationship

$$\hat{f}(a\phi + b\psi) = af\phi + bf\psi$$

where a and b are constants and ϕ and ψ are arbitrary functions. We shall use a circumflex to indicate that a symbol is to represent an operator. (Cf. Exercise 1.1.)

quantum mechanics.   The form of the operator of a physical quantity is much easier to decide, as is demonstrated in the next paragraph.

The following rules, by which the operators of physical quantities may be constructed, are included in postulate 2.

(i) If f is a coordinate q (for example, x, y, z, t), the operation is simply multiplication by q, that is,

$$\hat{q} \equiv q \tag{1.3}$$

(ii) If f is the momentum $\mathbf{p}$ which is conjugate[2] to the coordinate $\mathbf{q}$, that is

$$\mathbf{p} = M \cdot \dot{\mathbf{q}} \equiv M \cdot \frac{d\mathbf{q}}{dt}$$

then

$$\hat{p} = -i\hbar \frac{\partial}{\partial q} \tag{1.4}$$

where $\hbar$ is Planck's constant h divided by 2π.   In ordinary cartesian coordinates, for example,

$$p_x = m \cdot \dot{x}$$

and

$$\hat{p}_x = -i\hbar \frac{\partial}{\partial x} \tag{1.5}$$

(iii) If f is some function of coordinates and momenta, then $\hat{f}$ is the same function of the corresponding operators. A simple example is provided by the angular momentum $\mathbf{M}$, where for momentum about the z axis it can be shown that

$$M_z = x p_y - y p_x$$

and hence

$$\hat{M}_z = -i\hbar \left(x \frac{\partial}{\partial y} - y \frac{\partial}{\partial x} \right) \tag{1.6}$$

is the required operator.

If the function of coordinates and momenta is such that the order of factors is important, then the operator must be constructed in such a way that $\int \Psi_m^* \hat{f} \Psi_n \, d\tau = \int \Psi_n^* \hat{f}^* \Psi_m \, d\tau$, i.e., the operator is *Hermitian*.   Here Ψ_m and Ψ_n are two different wave functions of

---

[2] The classical definition of conjugate coordinates and momenta is $p_i = \partial L / \partial q_i$, where $L = T - V$ is called the Lagrangian function and is equal to the difference between the kinetic and potential energies of the system.

the system. It can be shown that this is required by the condition that the eigenvalues of f should be real. (Cf. Exercise 1.2.)

POSTULATE 3. If Ψ_1 and Ψ_2 represent possible states of a system, so does

$$\Psi_3 = a_1\Psi_1 + a_2\Psi_2 \tag{1.7}$$

where a_1 and a_2 are constants. This is known as the *principle of superposition*. If a physical quantity has the value f_1 in state Ψ_1 and the value f_2 in state Ψ_2, then in state Ψ_3 there is a probability $a_1^* \cdot a_1 = |a_1|^2$ of observing the value f_1 and a probability $|a_2|^2$ of observing the value f_2, where

$$|a_1|^2 + |a_2|^2 = 1 \tag{1.8}$$

This postulate is to some extent an expression of the fact that the wave function can be obtained as the solution of a linear differential equation (the Schroedinger equation). Each solution of this equation corresponds to a different state of the system, and any linear combination of functions which satisfy the equation must also be a possible solution. Hence the linear combination must also correspond to a possible state of the system.

1c IMMEDIATE CONSEQUENCES OF THE POSTULATES

Normalization of Wave Functions. According to postulate 1 the probability that the variables q_1 to q_n on which Ψ depends will lie in an element of configuration space $d\tau$ is equal to $\Psi^*\Psi\, d\tau$. Since the probability that they will lie *somewhere* in the whole configuration space is unity, we must have

$$\int \Psi^*\Psi\, d\tau = 1 \tag{1.9}$$

where the integral is taken over the whole of configuration space, i.e., over the whole accessible range of the variables q. When this equation holds the wave function is said to be *normalized*.

In practice we shall often work with unnormalized wave functions, only introducing the appropriate normalizing factor when it is desired to evaluate some physical quantity or as the last step in the calculation of an unknown wave function.

Matrix Elements.   Any integral of the type

$$f_{mn} = \int \Psi_m^* \hat{f} \Psi_n \, d\tau \tag{1.10}$$

is termed a "matrix element," on the grounds that we can imagine a matrix such as

$$F = (f_{mn}) = \begin{pmatrix} f_{11} & f_{12} & f_{13} & \cdots & f_{1r} \\ f_{21} & f_{22} & f_{23} & \cdots & f_{2r} \\ \cdot & \cdot & \cdot & \cdot & \cdot & \cdot \\ f_{r1} & f_{r2} & f_{r3} & \cdots & f_{rr} \end{pmatrix} \tag{1.11}$$

to be constructed by taking all of the r wave functions of a system in pairs.

Integrals of this type occur very frequently in quantum theory, but fortunately we are almost never required to evaluate them by direct integration.   Using this notation for matrix elements we can write the Hermitian requirement of postulate 2 in the convenient form

$$f_{mn} = f_{nm}^* \tag{1.12}$$

Orthogonality of Wave Functions.   Suppose that Ψ_m and Ψ_n are different eigenfunctions of an operator $\hat{f}$ having eigenvalues f_m and f_n, respectively.   Then, using the notation for matrix elements,

$$f_{mn} = \int \Psi_m^* \hat{f} \Psi_n \, d\tau$$
$$= f_n \int \Psi_m^* \Psi_n \, d\tau \tag{1.13}$$

since $\hat{f}\Psi_n = f_n\Psi_n$ by definition, and f_n is a constant.   Similarly,

$$f_{nm} = f_m \int \Psi_n^* \Psi_m \, d\tau \tag{1.14}$$

Taking the complex conjugate of both sides of this last equation, we find

$$f_{nm}^* = f_m \int \Psi_n \Psi_m^* \, d\tau \tag{1.15}$$

where we have used the fact that the physical quantity f must have a real eigenvalue, that is, $f_m^* = f_m$.   Subtracting (1.15) from (1.13) yields, in view of (1.12),

$$0 = (f_n - f_m) \int \Psi_m^* \Psi_n \, d\tau \tag{1.16}$$

(The order of factors in the integral is not significant once the operator has been taken out.) Hence if f_m and f_n are *different* eigen-values, we must have

$$0 = \int \Psi_m^* \Psi_n \, d\tau \qquad (1.17)$$

and the wave functions are said to be *orthogonal*.

If the different eigenfunctions correspond to the same eigenvalue, i.e., if they describe what is known as a degenerate state, they are no longer required to be orthogonal, but in this case it is always possible to form linear combinations of the eigenfunctions in such a way that these linear combinations are orthogonal to each other. For example, from two real functions Ψ_1 and Ψ_2 we can form $\phi_1 = (\Psi_1 + \Psi_2)$ and $\phi_2 = (\Psi_1 - \Psi_2)$, which are orthogonal provided Ψ_1 and Ψ_2 are normalized.

Expansion Theorems. Functions which are both orthogonal and normalized are said to be orthonormal, and all of the possible orthonormal wave functions of a system, when taken together, form an *orthonormal set*. The complete set of r orthonormal wave functions of a system behaves like a set of orthogonal unit vectors in a space of r dimensions, in that all members of a set are both linearly independent (cf. Exercise 1.4) and orthogonal to one another, but no other linearly independent or orthogonal functions are possible in this system. Any other wave function of the system can be completely expressed as a linear combination of the members of the orthonormal set. (It is helpful to visualize these statements for the case $r = 3$, where the three orthonormal wave functions are analogous to a set of three unit vectors directed along the x, y, and z coordinate axes, respectively.) Such a set of unit vectors can be constructed in a variety of ways, just as a set of coordinate axes can be constructed to have any desired orientation in space, and for convenience the wave functions are generally chosen to be eigenfunctions of some particular operator, $\hat{f}$ say. The matrix of $\hat{f}$ is then diagonal, that is,

$$f_{mn} = 0 \qquad \text{for } m \text{ not equal to } n$$

and $\qquad f_{mm} = f_m \qquad$ (Cf. equation 1.13)

These last two equations may conveniently be combined into the single expression

$$f_{mn} = \delta_n^m \cdot f_m \qquad (1.18)$$

where the Kronecker delta, δ_n^m, is defined to be unity when m equals n and zero otherwise.

It is often useful to expand an arbitrary wave function Ψ as a linear combination of the members of a set of orthonormal wave functions ϕ, i.e., to write

$$\Psi = a_1\phi_1 + a_2\phi_2 + \cdots + a_r\phi_r \qquad (1.19)$$

(This is equivalent to expressing an ordinary vector in r-dimensional space in terms of its components along the r coordinate axes.)

If we multiply both sides of equation (1.19) by ϕ_m^* and integrate with respect to $d\tau$, we obtain

$$\int \phi_m^*\Psi \, d\tau = a_m \qquad (1.20)$$

which is a useful expression for the values of the coefficients in the expansion.

If the ϕ_m are chosen to be eigenfunctions of an operator $\hat{f}$, we can let $\hat{f}$ operate on both sides of equation (1.19) and obtain

$$\hat{f}\Psi = a_1 f_1 \phi_1 + a_2 f_2 \phi_2 + \cdots + a_r f_r \phi_r \qquad (1.21)$$

Now if we multiply the left-hand side of (1.21) by Ψ^* and the right-hand side by $\sum_m a_m^* \phi_m^*$, which is the complex conjugate of the original expansion (1.19), and then integrate with respect to $d\tau$, we obtain

$$\int \Psi^*\hat{f}\Psi \, d\tau = |a_1|^2 f_1 + |a_2|^2 f_2 + \cdots + |a_r|^2 f_r \qquad (1.22)$$

Note that all of the terms on the right-hand side of the type $a_m^* \phi_m^* \cdot a_n f_n \phi_n$ must vanish when we integrate over $d\tau$ because of the orthogonality of the functions ϕ_m and ϕ_n. The only terms remaining are those for which $m = n$, and these are as given in (1.22).

By postulate 3 the right-hand side of (1.22) is equal to the *mean value*, $\bar{f}$, of the physical quantity f in the arbitrary state Ψ. Therefore we can write

$$\bar{f} = \int \Psi^*\hat{f}\Psi \, d\tau \qquad (1.23)$$

This is an important result, because this mean value of f is the value which we can expect to obtain by averaging the results of a large number of measurements on the arbitrary system.

Commutation of Operators. If a wave function Ψ is simultaneously an eigenfunction of two operators, $\hat{f}$ and $\hat{g}$, then

$$\hat{f}(\hat{g}\Psi) = \hat{f}(g\Psi) = g(\hat{f}\Psi) = gf\Psi = fg\Psi = \hat{g}(\hat{f}\Psi)$$

That is $$(\hat{f}\hat{g} - \hat{g}\hat{f})\Psi = 0 \tag{1.24}$$

which is conveniently written as

$$\{\hat{f}, \hat{g}\}\Psi = 0$$

or simply $$\{\hat{f}, \hat{g}\} = 0 \tag{1.25}$$

When this equation holds, the operators $\hat{f}$ and $\hat{g}$ are said to commute. Conversely it is true that if two operators commute, the corresponding physical quantities can simultaneously take definite values, i.e., a wave function can be an eigenfunction of both operators simultaneously. If, on the other hand, $\hat{f}$ and $\hat{g}$ do not commute, the product operator $(\hat{f}\hat{g})$ has no physical meaning, and the corresponding physical quantites cannot simultaneously take definite values. The fact that the operators $\hat{p}_x$ and $\hat{x}$ do not commute is the basis of the usual formulation of the Heisenberg uncertainty principle, namely

$$\Delta p_x \cdot \Delta x \geqslant \hbar \tag{1.26}$$

We are now in a position to appreciate one of the most fundamental differences between classical mechanics and quantum mechanics. This difference lies in the approach to the measurement of physical quantities. In classical mechanics it is tacitly assumed that when a property of a system is being measured the measuring apparatus can always be made sufficiently delicate so that it does not disturb the system in any way. In other words, the classical ideal is the detached observer who is not to be regarded as a part of the experiment. In quantum mechanics, on the other hand, any process of measurement consists of the interaction of the quantum mechanical system with some *classical* object, the "measuring apparatus," and the presence or absence of an observer is irrelevant. (The concept of measurement is therefore somewhat wider than in classical mechanics.) If an observer *is* present, the value of the physical quantity which is being measured is inferred from the observed change in the state of the measuring apparatus. It is apparent that in these circumstances the apparatus can never be made sufficiently delicate to leave the quantum mechanical system undisturbed, and that the measurement of one property may alter the state of the system so that the subsequent precise measurement of another property is impossible. Our mathematical analogue of the measuring process is the application of the appropriate linear operator to the wave function Ψ, and the situation where one measurement affects the result of another corresponds to the case where the operators in question do not commute.

It may be helpful to consider two examples of measurements of the type which occur in quantum mechanics. For the first example, suppose that a monoenergetic beam of electrons is directed onto a narrow slit. The position of an electron is fixed, i.e., measured, when it passes through the slit, and the velocities of the electrons in the emergent beam are no longer all the same, having been altered by interaction with the edges of the slit. The range of the velocity distribution increases as the slit is made more narrow, and ultimately a diffraction pattern is formed.

For the second example we consider a system which is in an excited state. Such a system can lose energy and return to the ground state by emission of electromagnetic radiation, which will ideally be emitted as a line of a single frequency v which is given by the Planck relationship $E = hv$. In fact, however, the operators for energy and time do not commute (see section 1d), and the emission "line" extends over a frequency range which is inversely proportional to the lifetime of the excited state. Here the measurement is the process of emission of a photon whose energy could, in principle, be determined by entirely classical means. In the absence of other factors, such as Doppler or pressure broadening, the natural width ΔE of a spectral line is related to the lifetime of the excited state Δt by the Heisenberg uncertainty principle, in the form

$$\Delta E \cdot \Delta t \sim \hbar$$

Some very short lifetimes (*ca.* 10^{-15} sec.) of excited states of atomic nuclei have actually been determined from measurements of natural linewidths with the aid of this form of the uncertainty principle.

In both of these examples the process of measurement consists essentially of forcing a physical quantity to take a definite value, the quantities in question being a space coordinate in the first case and a time (of emission) in the second.

1d THE HAMILTONIAN OPERATOR

In classical mechanics the total energy E of a system is given by Hamilton's function, which for a single particle takes the form

$$\mathcal{H} = \frac{1}{2m}(p_x^2 + p_y^2 + p_z^2) + V(x, y, z, t) \qquad (1.27)$$

We observe that the value of this function is given by the sum of the kinetic and potential energies of the system.

We can convert the function into an operator by applying the rules associated with postulate 2, and hence obtain

$$\hat{H} = -\frac{\hbar^2}{2m}\left(\frac{\partial^2}{\partial x^2} + \frac{\partial^2}{\partial y^2} + \frac{\partial^2}{\partial z^2}\right) + V(x, y, z, t) \qquad (1.28)$$

which can be written more succinctly in the form

$$\hat{H} = -\frac{\hbar^2}{2m} \cdot \nabla^2 + V(x, y, z, t) \qquad (1.29)$$

Here $\hat{H}$ is the *Hamiltonian operator*. For most purposes this is the most important of the quantum mechanical operators. The differential operator ∇^2 is known as the Laplacian operator. The potential energy term $V(x, y, z, t)$, being a function of coordinates alone, remains unchanged when we convert to the operator expression.

An operator which is very closely related to the Hamiltonian operator is the operator for differentiation of the wave function with respect to the time t, which we write in the form

$$i\hbar \, \partial/\partial t \qquad (1.30)$$

If the transition from quantum mechanics to classical mechanics is considered it can be shown[3] that the effect of this operator corresponds, in the classical limit, to multiplication of the wave function by Hamilton's function $\mathscr{H}$. Since the same is necessarily true of the Hamiltonian operator, we must have in general

$$\hat{H}\Psi = i\hbar \frac{\partial \Psi}{\partial t} \qquad (1.31)$$

We can now substitute the expression for $\hat{H}$, given in (1.29), into equation (1.31) and obtain

$$-\frac{\hbar^2}{2m} \nabla^2 \Psi + V(x, y, z, t)\Psi = i\hbar \frac{\partial \Psi}{\partial t} \qquad (1.32)$$

which is the *Schroedinger equation including the time*. It is often written simply in the form (1.31).

[3] For details see Landau and Lifshitz, *Quantum Mechanics*, London: Pergamon Press, 1959, p. 25.

Next let us suppose that we are dealing with a system for which the wave function is an eigenfunction of the Hamiltonian operator. In this case

$$\hat{H}\Psi = E\Psi \tag{1.33}$$

where E is the total energy of the system. Such a system is said to be in a stationary state of energy E. The operators $\hat{H}$ and $\hat{t}$ do not commute (as can most readily be seen from the fact that $\hat{H}$ is equal to $i\hbar\, \partial/\partial t$), and in a stationary state the time t does not take a definite value. The potential energy is therefore written simply as $V(x, y, z)$, independent of the time t.

If equations (1.31) and (1.33) are to be mutually consistent, the wave function Ψ must be of the form

$$\Psi(x, y, z, t) = \phi(x, y, z)e^{-Eit/\hbar} \tag{1.34}$$

where $\phi(x, y, z)$ is independent of time.

When we substitute this expression for Ψ into equation (1.32) we obtain

$$\frac{-\hbar^2}{2m}\nabla^2\phi e^{-Eit/\hbar} + V(x, y, z)\phi e^{-Eit/\hbar} = E\phi e^{-Eit/\hbar} \tag{1.35}$$

which immediately reduces to

$$\nabla^2\phi + \frac{2m}{\hbar^2}(E - V)\phi = 0 \tag{1.36}$$

This is the *Schroedinger equation for a stationary state.* It is often written simply in the form (1.33), where Ψ is understood to be independent of time.

We are usually interested in stationary states of systems, and the wave functions are then such that the matrix of $\hat{H}$ is diagonal. The importance of the Hamiltonian operator is such that most of our problems will consist of determining, for particular systems, the wave functions and energy levels which satisfy equation (1.36).

1e THE ANGULAR MOMENTUM

General Considerations. This section on angular momentum is rather long, but the results to be obtained are of such wide application that the effort to be spent on obtaining them is more than justified.   In addition it will provide some useful practice in the manipulation of operators and will also constitute an interesting example of how it is often possible (at least in quantum theory) to derive a great deal of valuable information from a relatively insignificant amount of starting material.   In this instance the basis of our calculations will be some rules for the commutation of angular momentum operators.

Classically the angular momentum about a point O is a vector, **M**, which is defined as the vector product of distance **r** from the point and linear momentum **p** (Fig. 1).

$$\mathbf{M} = \mathbf{r} \times \mathbf{p} = rp \sin \theta$$

Here by convention the vector **M** is directed outwards from O at right angles to the plane of the diagram.   In three dimensions a useful expression for our purposes is

$$\mathbf{M} = \begin{vmatrix} \mathbf{i} & \mathbf{j} & \mathbf{k} \\ x & y & z \\ p_x & p_y & p_z \end{vmatrix} \tag{1.37}$$

In equation (1.37), which is the standard determinant expression for a vector product, **i**, **j**, and **k** are unit vectors parallel to the x, y, and z coordinate axes, respectively, and (x, y, z), (p_x, p_y, p_z) are the components of the vectors **r** and **p**, respectively.[4]

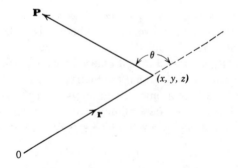

Fig. 1   Definition of a vector product. O is taken as the origin of coordinates.

[4] See, for example, B. Hague, *Vector Analysis*, 5th Edition, London: Methuen and Co. Ltd., 1957.

From postulate 2 we can immediately write down the corresponding operator expression, which is

$$\hat{M} = -i\hbar \begin{vmatrix} \mathbf{i} & \mathbf{j} & \mathbf{k} \\ x & y & z \\ \dfrac{\partial}{\partial x} & \dfrac{\partial}{\partial y} & \dfrac{\partial}{\partial z} \end{vmatrix} \tag{1.38}$$

and the components

$$\hat{M}_x = -i\hbar\left(y\frac{\partial}{\partial z} - z\frac{\partial}{\partial y}\right)$$

$$\hat{M}_y = -i\hbar\left(z\frac{\partial}{\partial x} - x\frac{\partial}{\partial z}\right) \tag{1.39}$$

$$\hat{M}_z = -i\hbar\left(x\frac{\partial}{\partial y} - y\frac{\partial}{\partial x}\right)$$

are obtained by expanding the determinant.

In practice the operator for total angular momentum $\hat{M}$ is less useful than the operator for the square of the total angular momentum, $\hat{M}^2$, where, since

$$M^2 = M_x^2 + M_y^2 + M_z^2$$

we have

$$\hat{M}^2 = \hat{M}_x^2 + \hat{M}_y^2 + \hat{M}_z^2 \tag{1.40}$$

Commutation Rules. Let us consider the effects of commuting these operators in pairs. For $\hat{M}_x$ and $\hat{M}_y$ we have

$$\hat{M}_x(\hat{M}_y\Psi) - \hat{M}_y(\hat{M}_x\Psi) \equiv \{\hat{M}_x, \hat{M}_y\}\Psi$$

$$= -\hbar^2\left(y\frac{\partial}{\partial z} - z\frac{\partial}{\partial y}\right)\left(z\frac{\partial\Psi}{\partial x} - x\frac{\partial\Psi}{\partial z}\right)$$

$$\qquad + \hbar^2\left(z\frac{\partial}{\partial x} - x\frac{\partial}{\partial z}\right)\left(y\frac{\partial\Psi}{\partial z} - z\frac{\partial\Psi}{\partial y}\right)$$

$$= \hbar^2\left(-yz\frac{\partial^2\Psi}{\partial z\,\partial x} - y\frac{\partial\Psi}{\partial x} + yx\frac{\partial^2\Psi}{\partial z^2} + z^2\frac{\partial^2\Psi}{\partial y\,\partial x} - zx\frac{\partial^2\Psi}{\partial y\,\partial z}\right.$$

$$\qquad \left. + zy\frac{\partial^2\Psi}{\partial x\,\partial z} - z^2\frac{\partial^2\Psi}{\partial x\,\partial y} - xy\frac{\partial^2\Psi}{\partial z^2} + xz\frac{\partial^2\Psi}{\partial z\,\partial y} + x\frac{\partial\Psi}{\partial y}\right)$$

$$= \hbar^2\left(x\frac{\partial\Psi}{\partial y} - y\frac{\partial\Psi}{\partial x}\right)$$

That is, we can write formally

$$(\hat{M}_x \hat{M}_y - \hat{M}_y \hat{M}_x) = \{\hat{M}_x, \hat{M}_y\} = i\hbar \hat{M}_z$$

and
$$\{\hat{M}_y, \hat{M}_z\} = i\hbar \hat{M}_x \qquad (1.41)$$

and
$$\{\hat{M}_z, \hat{M}_x\} = i\hbar \hat{M}_y$$

where the last two equations follow by symmetry.

These results indicate, since the operators do not commute, that it is only possible for one component of an angular momentum vector to possess a definite value.   Once the component of angular momentum in a particular direction is specified, the other components become uncertain.

For the operators $\hat{M}^2$ and $\hat{M}_z$ we have

$$\{\hat{M}^2, \hat{M}_z\} \equiv \hat{M}^2 \hat{M}_z - \hat{M}_z \hat{M}^2$$
$$= \{\hat{M}_x^2, \hat{M}_z\} + \{\hat{M}_y^2, \hat{M}_z\} \qquad (1.42)$$

where we have split $\hat{M}^2$ into its components according to (1.40) and have made use of the fact that $\{\hat{M}_z^2, \hat{M}_z\}$ is automatically equal to zero.   Now

$$\{\hat{M}_x^2, \hat{M}_z\} = \hat{M}_x \hat{M}_x \hat{M}_z - \hat{M}_z \hat{M}_x \hat{M}_x$$
$$= \hat{M}_x(\hat{M}_x \hat{M}_z - \hat{M}_z \hat{M}_x) - (\hat{M}_z \hat{M}_x - \hat{M}_x \hat{M}_z)\hat{M}_x \qquad (1.43)$$

where the order of the factors is important.

By (1.41) this becomes

$$\{\hat{M}_x^2, \hat{M}_z\} = -i\hbar(\hat{M}_x \hat{M}_y + \hat{M}_y \hat{M}_x) \qquad (1.44)$$

Similarly,

$$\{\hat{M}_y^2, \hat{M}_z\} = \hat{M}_y(\hat{M}_y \hat{M}_z - \hat{M}_z \hat{M}_y) - (\hat{M}_z \hat{M}_y - \hat{M}_y \hat{M}_z)\hat{M}_y$$
$$= +i\hbar(\hat{M}_y \hat{M}_x + \hat{M}_x \hat{M}_y) \qquad (1.45)$$

and so it follows that

$$\{\hat{M}^2, \hat{M}_z\} = 0 \qquad \text{i.e., the operators } do \text{ commute,}$$

and
$$\{\hat{M}^2, \hat{M}_y\} = 0 \qquad (1.46)$$

and
$$\{\hat{M}^2, \hat{M}_x\} = 0 \qquad \text{follow by symmetry.}$$

Therefore we have shown that a rotating system is characterized by two angular momentum eigenvalues, one for the square of the total angular momentum and one for the component of angular momentum in a particular direction, but no two components of the angular momentum can take definite values simultaneously.

Angular Momentum Eigenvalues.   Next we shall try to determine the relationship between the eigenvalues of $\hat{M}^2$ and of one of the components, $\hat{M}_z$ say, without considering the actual form of the angular momentum eigenfunctions.

Let us label the hypothetical eigenfunctions Y_{lm} where

$$\hat{M}^2 Y_{lm} = k_l Y_{lm} \tag{1.47}$$

and

$$\hat{M}_z Y_{lm} = k_m Y_{lm} \tag{1.48}$$

Here l and m are quantum numbers which specify the eigenvalues k_l and k_m. Operating on both sides of (1.48) with $\hat{M}_z$ gives

$$\hat{M}_z^2 Y_{lm} = k_m^2 Y_{lm}$$

so that

$$(\hat{M}_x^2 + \hat{M}_y^2) Y_{lm} = (k_l - k_m^2) Y_{lm} \tag{1.49}$$

Now since $(\hat{M}_x^2 + \hat{M}_y^2)$ is the operator of a physical quantity $(M_x^2 + M_y^2)$ which can never be negative, the same must be true of its eigenvalue in (1.49), that is,

$$k_l \geqslant k_m^2 \tag{1.50}$$

Next we consider the relations, which the reader can easily verify by expansion,

$$\hat{M}_z(\hat{M}_x + i\hat{M}_y) = (\hat{M}_x + i\hat{M}_y)(\hat{M}_z + \hbar)$$

and

$$\hat{M}_z(\hat{M}_x - i\hat{M}_y) = (\hat{M}_x - i\hat{M}_y)(\hat{M}_z - \hbar) \tag{1.51}$$

If these expressions are used to operate on Y_{lm}, we obtain

$$\hat{M}_z(\hat{M}_x + i\hat{M}_y) Y_{lm} = (\hat{M}_x + i\hat{M}_y)(k_m + \hbar) Y_{lm}$$

and

$$\hat{M}_z(\hat{M}_x - i\hat{M}_y) Y_{lm} = (\hat{M}_x - i\hat{M}_y)(k_m - \hbar) Y_{lm} \tag{1.52}$$

Therefore $[(\hat{M}_x + i\hat{M}_y) Y_{lm}]$ and $[(\hat{M}_x - i\hat{M}_y) Y_{lm}]$ are eigenfunctions of $\hat{M}_z$ having eigenvalues $(k_m + \hbar)$ and $(k_m - \hbar)$, respectively. Since $\hat{M}^2$ commutes with $\hat{M}_y$ and $\hat{M}_x$, these functions correspond to the same eigenvalue, k_l, of $\hat{M}^2$. Similarly we find that $[(\hat{M}_x \pm i\hat{M}_y)^2 \cdot Y_{lm}]$ is an eigenfunction of $\hat{M}_z$ with the eigenvalue $(k_m \pm 2\hbar)$, and so on. $((\hat{M}_x + i\hat{M}_y)$ and $(\hat{M}_x - i\hat{M}_y)$ are usually called the "raising" and "lowering" operators.)

Thus for each eigenvalue k_l of $\hat{M}^2$ we can use the operators $(\hat{M}_x \pm i\hat{M}_y)$ to generate a series of eigenfunctions which have the same eigenvalue of $\hat{M}^2$ but have different eigenvalues of $\hat{M}_z$. The eigenvalues of $\hat{M}_z$ are members of the series

$$k'_m, k'_m + \hbar, k'_m + 2\hbar, \cdots k''_m - 2\hbar, k''_m - \hbar, k''_m \qquad (1.53)$$

Here k'_m is the eigenvalue at the lower end and k''_m is the eigenvalue at the upper end of the series, the upper and lower limits of which are set by the condition (1.50). From the manner in which the members of the series are generated it follows that

$$k''_m = k'_m + n\hbar \qquad (1.54)$$

where n is some positive integer.

Let

$$\hat{M}_z Y'_{lm} = k'_m Y'_{lm} \qquad (1.55)$$

where k'_m is the *lowest* eigenvalue of $\hat{M}_z$. Then, in order to avoid generating a function with a still lower eigenvalue, we must have

$$(\hat{M}_x - i\hat{M}_y) Y'_{lm} = 0 \qquad (1.56)$$

Operating on this with $(\hat{M}_x + i\hat{M}_y)$ yields

$$\begin{aligned}
0 &= (\hat{M}_x^2 + \hat{M}_y^2 - i\hat{M}_x\hat{M}_y + i\hat{M}_y\hat{M}_x) Y'_{lm} \\
&= (\hat{M}^2 - \hat{M}_z^2 + \hbar\hat{M}_z) Y'_{lm} \\
&= (k_l - k'^2_m + \hbar k'_m) Y'_{lm} \qquad (1.57)
\end{aligned}$$

That is,

$$k_l = k'_m(k'_m - \hbar) \qquad (1.58)$$

since Y'_{lm} does not equal zero.

By the same sort of reasoning we can show that

$$k_l = k''_m(k''_m + \hbar) \qquad (1.59)$$

In combination with (1.54) these two equations require

$$-k'_m = +k''_m = \tfrac{1}{2}n\hbar = l\hbar \qquad (1.60)$$

where l is required to be an integer or half an integer. Hence, from (1.59)

$$k_l = l(l + 1)\hbar^2 \qquad (1.61)$$

and

$$k_m = m\hbar \qquad (1.62)$$

there be the same number, $(2l + 1)(2l' + 1)$, of states in the system considered as a whole, but there must be a $1 : 1$ correspondence between the new states and the old ones.

Since m, m', and M refer to components in the same direction, we must have simply

$$M = m + m' \tag{1.68}$$

The greatest possible value of M is now $(l + l')$, corresponding to $m = l$ and $m' = l'$, and this must also be the greatest possible value of L. The number of states of the whole system corresponding to this value of L is then $(2L + 1) = (2l + 2l' + 1)$.

The next greatest value of M is $(l + l' - 1)$, and when the parts of the system are considered separately it is seen that there are two states which correspond to this value: one state for $m = l$ and $m' = l' - 1$, and the other for $m = l - 1$ and $m' = l'$. In the combined system one of these states arises from $L = l + l'$, $M = L - 1$ and has already been counted. The other must arise from $L = l + l' - 1$, $M = L$. The number of states of the whole system corresponding to this value of L is then $(2l + 2l' - 1)$.

Continuing in this way we find that the number of states in the whole system is given by the sum of the series

$$(2l + 2l' + 1) + (2l + 2l' - 1) + (2l + 2l' - 3) + \cdots \tag{1.69}$$

If we assume that the series continues as far as the level for which $L = (l + l' - n)$, then it is easy to show, by summing the arithmetical progression (1.69), that n is equal to either $2l$ or $2l'$. Suppose $l > l'$. Then if we take n equal to $2l$ not only are some values of L negative, but it becomes apparent, on considering a definite example such as $l = 3$, $l' = 1$, that the number of states in the combined system is too large. Therefore the rule is that the series must continue as far as the term for which $L = l - l'$, where $l \geqslant l'$.

To summarize: if two angular momentum vectors are characterized by quantum numbers l, m and l', m', their resultant is characterized by quantum numbers L and M, where L takes the values

$$(l + l'), (l + l' - 1), (l + l' - 2), \cdots (l - l') \qquad [l \geqslant l'] \tag{1.70}$$

and M takes $2L + 1$ values for every value of L.

The angular momentum eigenvalues of the resultant are given as before by

$$\hat{M}^2 Y_{LM} = L(L + 1)\hbar^2 Y_{LM} \tag{1.71}$$

and $$\hat{M}_z Y_{LM} = M\hbar Y_{LM} \tag{1.72}$$

It is sometimes necessary to distinguish the resultants of different types of angular momentum vectors, for example, the vectors **L** and **S** of the orbital and spin momentum for electrons in an atom. In this case we label the z components M_L and M_S, respectively, while the resultant of **L** and **S**, usually written **J**, has a z component M_J. We shall meet some examples of this in Chapter 6.

EXERCISES

1.1 (*a*) Which of the following are linear operators? (i) d/dx; (ii) ∇^2; (iii) Multiply by a constant; (iv) Add a constant; (v) Take the square root; (vi) Leave it as it is.

(*b*) Find whether any of the following functions are eigenfunctions of any of the operators in (*a*): (i) x^a; (ii) e^{ax}; (iii) $\log ax$; (iv) $\cos ax$; (v) $\cos ax + i \sin ax$. (Here a is a constant.)

(*c*) Given a wave function $\psi = \sqrt{2/a} \cdot \sin(\pi x/a)$ for a particle of mass m which is confined within the region $0 < x < a$, calculate: (i) whether the wave function is normalized; (ii) the momentum p_x; (iii) the kinetic energy $T = p_x^2/2m$; (iv) the probability of finding the particle outside the region $0.25a < x < 0.75a$.

1.2 Satisfy yourself that the derivation of equation (1.23) for the mean value $\bar{f}$ of a physical quantity does not depend on the operator $\hat{f}$ being Hermitian.   Use (1.23) to show that $\hat{f}$ must be Hermitian if $\bar{f}$ is to be real.

1.3 (*a*) Prove that $\hat{f}\Psi_n = \sum_m f_{mn}\Psi_m$.   (Expand $\hat{f}\Psi_n$ in terms of the set of functions Ψ_m, and find the coefficients as before.)

(*b*) Using the equation $\hat{g}\Psi_n = \sum_k g_{kn}\Psi_k$, work out the expression for $(\hat{f}\hat{g})\Psi_n$.   Hence verify that the product of two matrices $F = (f_{mn})$ and $G = (g_{mn})$ is given by the matrix multiplication rule, that is, $(FG)_{kn} = \sum_m f_{km}g_{mn}$ where FG is the matrix of the operator $(\hat{f}\hat{g})$.

1.4 Prove that the expansion of Ψ in terms of an orthonormal set of wave functions does not contain Ψ_m if Ψ and Ψ_m are orthogonal, i.e., show that Ψ and Ψ_m are linearly independent in this case.

1.5 In a commonly used notation due to Dirac, a wave function (vector) is written, for example, as $|\Psi_m\rangle$ or more simply as $|m\rangle$. (This is pronounced *ket m*). The complex conjugate wave function is written $\langle m|$. (Pronounced *bra m*, a "scalar product" such as $\langle m|n\rangle$ is pronounced *bracket mn*.) A matrix element is then written, for example, as $\langle m|H|n\rangle$. Verify, for m and n taking the values 0, 1, 2, 3, that when the complete matrix of $\langle m|H|n\rangle$ is being formed $\langle m|$ behaves as a row matrix (vector) and $|n\rangle$ behaves as a column matrix (or vector). (Cf. the matrix multiplication rule. We shall not use this notation, but it occurs very frequently in the literature.)

1.6 Commute the following pairs of operators: (i) $\hat{P}_x$ and $\hat{P}_y$; (ii) $\hat{P}_x$ and $\hat{x}$; (iii) $\hat{P}_x$ and $\hat{Z}$; (iv) $\hat{H}$ and $\hat{t}$; (v) $\hat{M}_z$ and $\hat{M}_y$; (vi) $\hat{M}^2$ and $\hat{M}_y$; (vii) $\hat{M}_z$ and $\hat{Z}$. Put into words your conclusions with regard to the possibility of simultaneously observing the corresponding pairs of physical quantities.

1.7 Draw diagrams to illustrate the addition of angular momentum vectors of magnitude $\sqrt{6}\hbar$ and $\sqrt{2}\hbar$. Show how precession occurs in the manner of Fig. 2. (Assume that there is an external field directed along the z axis.)

1.8 (*a*) Compare the ratios $\Delta\nu/\nu$, where $\Delta\nu$ is the natural line width and ν the frequency, for visible light of wavelength 5896 Å emitted by an excited sodium atom ($\tau = 1.6 \times 10^{-8}$ sec.) and for the γ rays of energy 14.4 kev and 93 kev emitted by excited nuclei of Fe^{57} ($\tau = 1.4 \times 10^{-7}$ sec.) and Zn^{67} ($\tau = 1.4 \times 10^{-5}$ sec.), respectively. Here τ stands for the mean lifetime of the excited state.

(*b*) Calculate the relative velocity of source and observer for which the Doppler shift in frequency is equal to the natural line width in each case. [In the Doppler effect $\nu' = \nu(1 + v/c)$, where $\nu' =$ shifted frequency, $v =$ relative velocity, $c =$ velocity of light (3×10^{10} cm./sec.)]

2

Solutions of the Schroedinger Equation for Some Simple Systems

2a INTRODUCTION

In this chapter we are concerned with the simplest of chemical systems, a single molecule in the gas phase. Some aspects of this system can be treated exactly in quantum mechanics by solving the Schroedinger equation, and we shall deal with these aspects here. The results to be obtained are of considerable importance in spectroscopy and statistical thermodynamics and also provide starting points for a number of approximate calculations on more complex systems.

The complete wave function for a gaseous molecule is necessarily quite complicated, since it must account for translation of the molecule as a whole in three dimensions, rotation of the molecule about different axes through its center of gravity, vibrations relative to one another of the atoms comprizing the molecule, the bonding forces which hold the atoms together, the electronic structures of the individual atoms, and finally the short-range forces between the constituents of the atomic nuclei. Fortunately it is usually possible to separate the components of the system and study them one at a time, i.e., it is a good approximation to write

$$\Psi = \Psi_1 \cdot \Psi_2 \cdot \Psi_3 \cdot \Psi_4 \cdot \Psi_5 \cdot \Psi_6 \tag{2.1}$$

and
$$\hat{H} = \hat{H}_1 + \hat{H}_2 + \hat{H}_3 + \hat{H}_4 + \hat{H}_5 + \hat{H}_6 \tag{2.2}$$

where
$$\hat{H}_1 \Psi_1 = E_1 \Psi_1, \text{ etc.} \tag{2.3}$$

24

Here Ψ and $\hat{H}$ refer to the entire system and the subscripts 1, 2, $\cdots$ 6 designate translational, rotational, vibrational, bonding, atomic, and nuclear contributions, respectively. The total energy of the system is then given by

$$E = E_1 + E_2 + E_3 + E_4 + E_5 + E_6 \qquad (2.4)$$

We shall endeavor to obtain some insight into the nature of Ψ_1, Ψ_2, Ψ_3, and Ψ_5 by considering certain idealized systems, namely the particle in a box, the rigid rotator, the harmonic oscillator, and the hydrogen atom. More complicated atoms than hydrogen and some features of chemical bonding will be discussed in Chapters 5 and 6. The current theory of the atomic nucleus will be left (regretfully) to the physicists.

2b THE PARTICLE IN A BOX

Our discussion will include the case of a free particle, since we can let the dimensions of the box tend to infinity. We consider a particle of mass m contained in a box of sides a, b, and c and volume $\tau = abc$. One corner of the box is taken as the origin of a cartesian coordinate system. The potential energy of the particle is assumed to be zero within the box and infinite everywhere else. This means that the wave function is zero outside the box and therefore, because Ψ is a continuous function, it must also tend to zero as the boundary is approached from within the box. We shall solve the Schroedinger equation for a stationary state, (1.36), to find the wave functions and energy levels of this system.

For $V = 0$ the Schroedinger equation is

$$\frac{\partial^2 \Psi}{\partial x^2} + \frac{\partial^2 \Psi}{\partial y^2} + \frac{\partial^2 \Psi}{\partial z^2} + \frac{2mE}{\hbar^2}\Psi = 0 \qquad (2.5)$$

The variables in this equation are separable, i.e., we can substitute

$$E = E_x + E_y + E_z \qquad (2.6)$$

and

$$\Psi = X(x) \cdot Y(y) \cdot Z(z) \qquad (2.7)$$

whereupon equation (2.5) can be split into three similar equations, of which a typical one is

$$\frac{d^2 X}{dx^2} + \frac{2mE_x X}{\hbar^2} = 0 \qquad (2.8)$$

The general solution[1] of equation (2.8) is

$$X = Ae^{ix(2mE_z)^{1/2}/\hbar} + Be^{-ix(2mE_z)^{1/2}/\hbar}$$

An alternative form which is more suitable for our purpose is

$$X = C \cdot \cos \frac{(2mE_x)^{1/2}x}{\hbar} + D \cdot \sin \frac{(2mE_x)^{1/2}x}{\hbar} \tag{2.9}$$

Since X is zero at $x = 0$ the constant C must be zero. The condition that X is zero at $x = a$ requires

$$E_x = n_x^2 \cdot \pi^2 \cdot \hbar^2 / 2ma^2 \tag{2.10}$$

where n_x is an integer. These wave functions and energy levels, for the first few values of n_x, are illustrated in Fig. 3.

Two important facts are immediately apparent. The first is that the introduction of boundary conditions has caused the energy to be quantized, i.e., to be restricted to certain discrete values which are specified by the quantum number n_x. The size of the energy quanta is inversely proportional to the square of the size of the box. The second fact is that n_x is never zero if there is to be any probability of finding the particle in the box, i.e., the particle can never possess less than the *zero-point energy*, which is given by equation (2.10) with $n_x = 1$.

We now have to find the value of D for which the wave function X is normalized. The normalizing condition is

$$\int_{-\infty}^{\infty} X^2 \, dx = \int_0^a X^2 \, dx = 1$$

that is,

$$D^2 \int_0^a \sin^2 (n_x \pi x/a) \, dx = 1$$

and so

$$D = \sqrt{2/a} \tag{2.11}$$

Combining these results with the similar ones for Y and Z, we obtain finally

$$\Psi = (8/\tau)^{1/2} \sin (n_x \pi x/a) \sin (n_y \pi y/b) \sin (n_z \pi z/c) \tag{2.12}$$

and

$$E = \frac{\pi^2 \hbar^2}{2m} \left(\frac{n_x^2}{a^2} + \frac{n_y^2}{b^2} + \frac{n_z^2}{c^2} \right) \tag{2.13}$$

[1] To solve the equation we substitute $X = Ae^{ux}$ in (2.8) and obtain $u^2 + 2mE_x/\hbar^2 = 0$, so that $u = \pm \sqrt{2mE_x}/\hbar$. This gives two possible solutions, and the general solution is a superposition of these. To change this solution into the form (2.9), we use $e^{ix} = \cos x + i \sin x$.

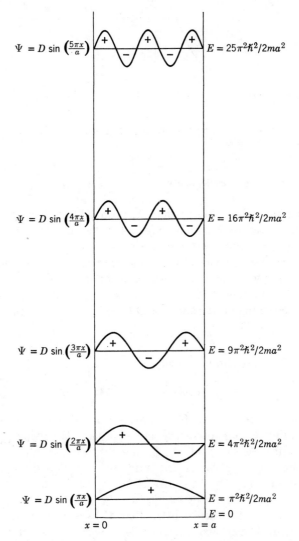

Fig. 3 The first five wave functions and energy levels for a particle in a one-dimensional box of length a.

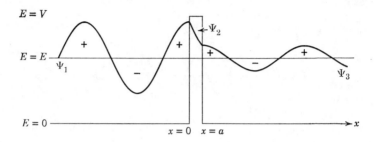

Fig. 4   Penetration of a rectangular potential barrier.   (The real parts of Ψ_1 and Ψ_3 are plotted.)

$$\Psi_1 = Ae^{ikx} + Be^{-ikx}$$
$$\Psi_2 = Ce^{k'z} + De^{-k'z}$$
$$\Psi_3 = Ee^{ikx}$$

We note that with a free particle the probability of finding it at a point tends to zero as τ tends to infinity and that the energy becomes completely unquantized at the same time.

Next let us briefly consider the case where there is a *potential barrier* of a finite height V and of width a, as shown in Fig. 4.

By the same reasoning as was used before the general solution of the Schroedinger equation for this case is

$$\Psi = Ae^{ikx} + Be^{-ikx} \tag{2.14}$$

where A and B are constants and

$$k = \sqrt{2m(E - V)/\hbar^2} \tag{2.15}$$

The real part of this wave function is drawn on the left side of the barrier in Fig. 4. Suppose we begin with a particle moving toward the barrier from left to right, i.e., having positive momentum. The momentum p_x is positive for the term Ae^{ikx}, so that the incident particle is described by this part of the wave function, while a reflected particle is described by the term Be^{-ikx}.

If the energy of the particle is less than V, as is the case when the particle is *inside* the barrier (classically this case does not exist!), the quantity ik becomes entirely real, and the wave function decays exponentially with increasing x, as shown in the figure. To the right of the barrier the wave function reappears in its original form, but attenuated by a factor $e^{k'a}$, where a is the thickness of the barrier, and

$$k' = \sqrt{2m(V - E)/\hbar^2} \qquad (2.16)$$

This is an illustration of the "tunnel effect," in which there is a finite probability that a particle will penetrate a barrier which it has too little energy to surmount. A closely related phenomenon is that the wave function of a particle may be partially reflected by a barrier which classically should not be high enough to affect it. This point is demonstrated in Exercise 2.1.

2c THE RIGID ROTATOR

The simplest rigid rotator consists of two point masses which are held apart at a fixed distance by a perfectly rigid and weightless bond. This ideal is approached quite closely by a diatomic molecule in its ground vibrational state. Classically the angular momentum and kinetic energy of such a system are given by

$$\mathbf{M} = I\boldsymbol{\omega}$$

and
$$T = E - V$$

$$= \tfrac{1}{2}I\omega^2$$

$$= M^2/2I \qquad (2.17)$$

where
$$I = \frac{m_1 \cdot m_2}{m_1 + m_2} a^2$$

$$= \mu a^2 \qquad (2.18)$$

Here I is the moment of inertia, a the bond length, $\boldsymbol{\omega}$ the angular velocity, and μ the reduced mass of the system. The system is mathematically equivalent to a particle of mass μ moving on the surface of a sphere of radius a. If the rotator is not acted on by any external forces, we can take V to be equal to zero on the surface of the sphere, and the Schroedinger equation becomes

$$\nabla^2\Psi + 2\mu E\Psi/\hbar^2 = 0 \qquad (2.19)$$

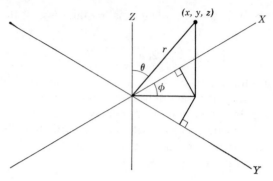

Fig. 5   Spherical polar coordinates.

We have shown in Chapter 1 that angular momentum eigen-functions and eigenvalues obey the relations

$$\hat{M}^2 \cdot Y_{lm} = l(l + 1)\hbar^2 \cdot Y_{lm}$$

and $$\hat{M}_z \cdot Y_{lm} = m\hbar \cdot Y_{lm} \quad (m = l, l - 1, \cdots - l)$$

Since the angular momentum operators commute with the Hamiltonian, the functions Y_{lm} are also eigenfunctions of the Hamiltonian operator.[2]   Our problem now is to determine the mathematical form of these functions, which are known as *spherical harmonics*.

In spherical polar coordinates r, θ, ϕ (Fig. 5), the Laplacian operator ∇^2 takes the form[3]

$$\frac{1}{r^2} \frac{\partial}{\partial r} \left(r^2 \frac{\partial}{\partial r} \right) + \frac{1}{r^2 \sin^2 \theta} \frac{\partial}{\partial \theta} \sin \theta \frac{\partial}{\partial \theta} + \frac{1}{r^2 \sin^2 \theta} \frac{\partial^2}{\partial \phi^2} \quad (2.20)$$

where $x = r \sin \theta \cos \phi$, $y = r \sin \theta \sin \phi$, $z = r \cos \theta$, and $d\tau = r^2 \sin \theta \cdot dr \, d\theta \, d\phi$.

[2] This is true provided that the potential V is invariant under rotation; otherwise angular momentum is not conserved, even in classical mechanics. The operators of physical quantities which are conserved in classical mechanics (e.g., linear and angular momentum in an isolated system) always commute with $\hat{H}$, and this is used as a definition of conserved quantities in quantum mechanics.

[3] For a fairly simple account of the transformation of ∇^2 into various co-ordinate systems see Eyring, Walter, and Kimball, *Quantum Chemistry*, Appendix 3.   New York: Wiley, 1944 (cf. also Exercise 2.2).

For the present system we put $r = a$, a constant, and the first term in ∇^2 vanishes. The Schroedinger equation now becomes

$$\frac{1}{\sin\theta}\frac{\partial}{\partial\theta}\left(\sin\theta\frac{\partial\Psi}{\partial\theta}\right) + \frac{1}{\sin^2\theta}\frac{\partial^2\Psi}{\partial\phi^2} + \frac{2IE\Psi}{\hbar^2} = 0 \qquad (2.21)$$

Before we solve equation (2.21) it will be instructive to also transform the operator $\hat{M}^2$ into spherical polar coordinates. From Fig. 5 we see that

$$\tan\phi = y/x \qquad (2.22)$$

$$\cos\theta = z/(x^2 + y^2 + z^2)^{1/2} \qquad (2.23)$$

$$r = (x^2 + y^2 + z^2)^{1/2} \qquad (2.24)$$

Hence, using

$$\frac{\partial}{\partial x} = \frac{\partial r}{\partial x}\frac{\partial}{\partial r} + \frac{\partial\theta}{\partial x}\frac{\partial}{\partial\theta} + \frac{\partial\phi}{\partial x}\frac{\partial}{\partial\phi} \qquad (2.25)$$

and the similar expressions for $\partial/\partial y$ and $\partial/\partial z$, we find

$$\frac{\partial}{\partial x} = \sin\theta\cos\phi\frac{\partial}{\partial r} + \frac{1}{r}\cos\theta\cos\phi\frac{\partial}{\partial\theta} - \frac{1}{r}\frac{\sin\phi}{\sin\theta}\frac{\partial}{\partial\phi} \qquad (2.26)$$

$$\frac{\partial}{\partial y} = \sin\theta\sin\phi\frac{\partial}{\partial r} + \frac{1}{r}\cos\theta\sin\phi\frac{\partial}{\partial\theta} + \frac{1}{r}\frac{\cos\phi}{\sin\theta}\frac{\partial}{\partial\phi} \qquad (2.27)$$

$$\frac{\partial}{\partial z} = \cos\theta\frac{\partial}{\partial r} - \frac{1}{r}\sin\theta\frac{\partial}{\partial\theta} \qquad (2.28)$$

If these results are introduced into the formulas for $\hat{M}_x$, $\hat{M}_y$, $\hat{M}_z$, and the resulting expressions are substituted in

$$\hat{M}^2 = (\hat{M}_x^2 + \hat{M}_y^2 + \hat{M}_z^2)$$

we obtain, after a good deal of manipulation,

$$\hat{M}^2 = -\hbar^2\left\{\frac{1}{\sin\theta}\frac{\partial}{\partial\theta}\left(\sin\theta\frac{\partial}{\partial\theta}\right) + \frac{1}{\sin^2\theta}\frac{\partial^2}{\partial\phi^2}\right\} \qquad (2.29)$$

Now $\hat{M}^2 Y_{lm} = l(l+1)\hbar^2 Y_{lm}$ so we must have

$$\frac{1}{\sin\theta}\frac{\partial}{\partial\theta}\left(\sin\theta\frac{\partial Y_{lm}}{\partial\theta}\right) + \frac{1}{\sin^2\theta}\frac{\partial^2 Y_{lm}}{\partial\phi^2} + l(l+1)Y_{lm} = 0 \qquad (2.30)$$

Comparison of (2.21) and (2.30) now reveals that

$$E_l = l(l+1)\hbar^2/2I \qquad (2.31)$$

where the E_l are energy eigenvalues for the rigid rotator. We note that the classical relation (2.17) also holds in quantum mechanics. States with the same value of l but different values of the second quantum number m are seen to be degenerate, i.e., to have the same energy. (This degeneracy can be removed by the interaction of **M** with an external field, as in the Zeeman and Stark effects.)

To solve equation (2.21) we make the substitution

$$\Psi \equiv Y_{lm} = Th(\theta) \cdot F(\phi) \tag{2.32}$$

and obtain

$$-\frac{1}{F}\frac{\partial^2 F}{\partial \phi^2} = \frac{\sin\theta}{Th}\frac{\partial}{\partial\theta}\left(\sin\theta\frac{\partial Th}{\partial\theta}\right) + \frac{2IE\sin^2\theta}{\hbar^2} \tag{2.33}$$

Now if we vary ϕ while keeping θ constant, we see that the right-hand side of (2.33) must be equal to a constant, which we will call m^2 in anticipation of the final result. Thus we obtain two separate equations

$$\frac{d^2 F}{d\phi^2} = -m^2 F \tag{2.34}$$

and

$$\frac{1}{\sin\theta}\frac{\partial}{\partial\theta}\left(\sin\theta\frac{\partial Th}{\partial\theta}\right) - \frac{m^2 Th}{\sin^2\theta} + l(l+1)Th = 0 \tag{2.35}$$

where we have made use of (2.31) to substitute for E in the second equation.

The solution of equation (2.34) is simply

$$F(\phi) = \text{constant} \times e^{im\phi} \tag{2.36}$$

where, since F must be single-valued, we require

$$e^{im\phi} = e^{im(\phi + 2\pi)} \tag{2.37}$$

That is,

$$e^{2\pi m i} = 1 \tag{2.38}$$

This is true only if m is zero or a positive or negative integer. When F is normalized by integrating over $d\phi$ from 0 to 2π, the constant in (2.36) becomes $1/(2\pi)^{1/2}$, so we have finally

$$F(\phi) = (2\pi)^{-1/2}e^{im\phi} \qquad (m = 0, \pm 1, \pm 2, \cdots) \tag{2.39}$$

The next step is to solve equation (2.35). If we substitute $x = \cos\theta$, we obtain

$$(1 - x^2)\frac{d^2 Th}{dx^2} - 2x\frac{dTh}{dx} + \left[l(l+1) - \frac{m^2}{1 - x^2}\right]Th = 0 \tag{2.40}$$

This equation is now in a standard mathematical form. Its solutions are known as Associated Legendre Polynomials, and in order to see how they arise we first consider equation (2.41).

$$(1 - x^2)\frac{dy}{dx} + 2lxy = 0 \tag{2.41}$$

(These details may be omitted at a first reading. Begin again after equation 2.50.)

Here the variables are immediately separable, and integration yields

$$y = c(1 - x^2)^l \tag{2.42}$$

If we differentiate equation (2.42) a total of $(l + 1)$ times (l is therefore required to be an integer from now on) the result is

$$(1 - x^2)\frac{d^{l+2}y}{dx^{l+2}} - 2x\frac{d^{l+1}y}{dx^{l+1}} + l(l + 1)\frac{d^l y}{dx^l} = 0 \tag{2.43}$$

Now if we put

$$z = c\frac{d^l}{dx^l}(1 - x^2)^l \tag{2.44}$$

we obtain

$$(1 - x^2)\frac{d^2 z}{dx^2} - 2x\frac{dz}{dx} + l(l + 1)z = 0 \tag{2.45}$$

Equation (2.45) is known as Legendre's equation, and the particular solution of this equation which is given by

$$z = P_l(x) = \frac{1}{2^l \cdot l!} \cdot \frac{d^l}{dx^l}(x^2 - 1)^l \tag{2.46}$$

is called the Legendre Polynomial of degree l. $P_0(x)$ is defined to be unity.

General solutions of Legendre's equation may be found in the form of infinite series which reduce to polynomials when l is an integer. These series can be made to converge for $|x| \neq 1$, but as we are later going to put x equal to cos θ, we shall only be interested in the solutions which are given by (2.46).

If equation (2.45) is differentiated a further m times, the result is

$$(1 - x^2)\frac{d^{m+2}z}{dx^{m+2}} - (2m + 1)x\frac{d^{m+1}z}{dx^{m+1}} + (m + l + 1)(l - m)\frac{d^m z}{dx^m} = 0 \tag{2.47}$$

We transform this equation in stages, putting first $d^m z/dx^m = u$, so that

$$(1 - x^2)\frac{d^2u}{dx^2} - (2m + 1)x\frac{du}{dx} + (m + l + 1)(l - m)u = 0 \quad (2.48)$$

Now we put $u = v(1 - x^2)^{-m/2}$ and obtain

$$(1 - x^2)\frac{d^2v}{dx^2} - 2x\frac{dv}{dx} + \left[l(l + 1) - \frac{m^2}{1 + x^2}\right]v = 0 \quad (2.49)$$

The particular solution of (2.49) which is given by

$$v = P_l^m(x) = (1 - x^2)^{m/2}\cdot\frac{d^m}{dx^m}[P_l(x)] \quad (2.50)$$

is known as the Associated Legendre Polynomial of degree l and order m.   We note that l and m are both integers and that m cannot be greater than l or the polynomial becomes zero.

Equations (2.40) and (2.49) are seen to be identical, so that the expression for $Th(\theta)$ is

$$Th(\theta) = N\cdot P_l^{|m|}(\cos\theta) \quad (2.51)$$

where N is a normalizing factor.   Here we have written $|m|$ to accommodate the fact that our quantum number m can be positive or negative. $(P_l^m(x)$ is not normally defined for negative m.)

It can be shown[4] that the normalizing factor N is given by

$$N = \sqrt{\frac{2l + 1}{2}\frac{(l - |m|)!}{(l + |m|)!}} \quad (2.52)$$

Hence our complete expression for the eigenfunctions of the rigid rotator, i.e., for the spherical harmonics, Y_{lm}, is

$$Y_{lm}(\theta, \phi) = \sqrt{\frac{2l + 1}{4\pi}\frac{(l - |m|)!}{(l + |m|)!}}\cdot P_l^{|m|}(\cos\theta)\cdot e^{im\phi} \quad (2.53)$$

Molecules in general have several moments of inertia, and the expressions for the eigenfunctions and energy levels are correspondingly more complex.   For further information about rotational energy levels of molecules the student should consult texts on molecular spectroscopy.[5]

---

[4] See Eyring, Walter, and Kimball, *op. cit.*, Chapter 4.
[5] For example, Townes and Shawlow, *Microwave Spectroscopy*, New York: McGraw-Hill, 1955.

2d THE HARMONIC OSCILLATOR

In classical mechanics a particle executes simple harmonic motion when it is subject to a potential

$$V = \tfrac{1}{2}kx^2 \tag{2.54}$$

where x is the displacement of the particle from its equilibrium position. The restoring force acting on the particle is $-kx$, and the equation of motion for the system is

$$m\frac{d^2x}{dt^2} + kx = 0 \tag{2.55}$$

This equation is satisfied by

$$x = a \cos \omega(t - t_0) \tag{2.56}$$

where a and t_0 are constants and

$$\omega = 2\pi\nu = (k/m)^{1/2} \tag{2.57}$$

is the (circular) frequency of the motion.

Provided the amplitude a is not too large, the harmonic oscillator is an excellent model for a vibrating diatomic molecule. The reduced mass $\mu = m_1 m_2/(m_1 + m_2)$ takes the place of m, and k becomes the force constant of the bond joining the atoms.

The eigenfunctions and eigenvalues of the harmonic oscillator were calculated by Heisenberg in 1925, by a matrix method, before the advent of the Schroedinger equation. It is instructive to follow through this calculation[6]; however, for the present we shall again proceed by solving the Schroedinger equation.

The classical Hamiltonian is

$$\mathscr{H} = \frac{p_x^2}{2\mu} + \frac{kx^2}{2}$$

and so the Hamiltonian operator is

$$\hat{H} = \frac{-\hbar^2}{2\mu}\frac{\partial^2}{\partial x^2} + \frac{kx^2}{2} \tag{2.58}$$

[6] See Landau and Lifshitz, *Quantum Mechanics*, London: Pergamon Press, 1959, p. 64. (Cf. also Exercise 2.5.)

by postulate 2. The Schroedinger equation therefore becomes

$$\frac{d^2\Psi}{dx^2} + \frac{2\mu}{\hbar^2}\left(E - \frac{kx^2}{2}\right)\Psi = 0 \tag{2.59}$$

We use the abbreviations $\alpha = 2\mu E/\hbar^2$, $\beta = \mu\omega/\hbar$ and substitute $\xi = x\sqrt{\beta}$. This gives

$$\frac{d^2\Psi}{d\xi^2} + \left(\frac{\alpha}{\beta} - \xi^2\right)\Psi = 0 \tag{2.60}$$

Now if ξ^2 is very large, equation (2.60) has the approximate solution

$$\Psi = C \cdot e^{\pm\xi^2/2} \tag{2.61}$$

as may be verified by substituting this expression into equation (2.60) and neglecting both α/β and 1 in comparison with ξ^2. The solution with the positive sign is not acceptable because it tends to infinity as ξ tends to infinity, but the solution with the negative sign will allow Ψ to remain finite, and so we try to obtain an exact solution in the form

$$\Psi = u \cdot e^{-\xi^2/2} \tag{2.62}$$

where u is some function of ξ. We find, on carrying out the appropriate substitution, that

$$\frac{d^2u}{d\xi^2} - 2\xi\frac{du}{d\xi} + 2nu = 0 \tag{2.63}$$

where we have now written $2n$ in place of $(\alpha/\beta - 1)$.

Equation (2.63) is a standard form which is known as Hermite's equation. The general solution of this equation is a superposition of two infinite series which converge for $|\xi| < 1$, but we are again interested only in the particular solutions which result when one of the series reduces to a polynomial, because there is then no restriction on the value of ξ. These solutions occur when n is an integer, and they are known as Hermite polynomials, $H_n(\xi)$, where

$$H_n(\xi) = (2\xi)^n - \frac{n(n-1)}{1!}(2\xi)^{n-2} + \cdots$$

$$+ (-1)^r\frac{n(n-1)\cdots(n-2r+1)}{r!}(2\xi)^{n-2r} + \cdots \tag{2.64}$$

The series (2.64) is continued as far as the last term for which $(n-2r)$ is not negative, i.e., as far as the constant term if there is one, or as far as the lowest positive power of ξ.

Therefore the eigenfunctions of the harmonic oscillator are given by

$$\Psi_n(\xi) = N \cdot H_n(\xi) \cdot e^{-\xi^2/2} \qquad (2.65)$$

where N is a normalizing factor. A few of these wave functions are shown graphically in Fig. 6.

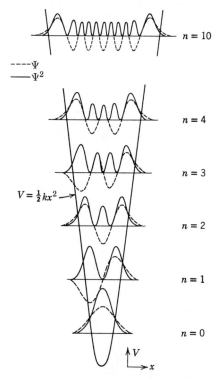

Fig. 6 Eigenfunctions, and probability functions, of a harmonic oscillator (from G. Herzberg, "Infrared and Raman Spectra of Polyatomic Molecules," *Molecular Spectra and Molecular Structure*, Vol. II, Princeton: D. Van Nostrand, 1945).

After normalization (see footnote 4) equation (2.65) becomes

$$\Psi_n(\xi) = \left(\frac{\sqrt{\beta/\pi}}{2^n \cdot n!}\right)^{1/2} \cdot H_n(\xi) \cdot e^{-\xi^2/2} \qquad (2.66)$$

where $\qquad \xi = \sqrt{\beta} \cdot x$

The wave functions in Fig. 6 are seen to be quite similar to those of a particle in a one-dimensional box, especially for small values of n.

To find the energy eigenvalues we return to the condition $(\alpha/\beta - 1) = 2n$, where n is required to be an integer for the solution to (2.63) to be a polynomial rather than an infinite series.   Using the definitions $\alpha = 2\mu E/\hbar^2$ and $\beta = \mu\omega/\hbar$, we find

$$E = (n + \tfrac{1}{2})\hbar\omega \qquad (2.67)$$

$$= (n + \tfrac{1}{2})h\nu \qquad (2.68)$$

This result is probably well known to the reader.   Once more the energy is quantized, and we also note the presence of a zero-point energy, which in this case is equal to $\tfrac{1}{2}h\nu$.

2e   THE HYDROGEN ATOM

In the hydrogen atom we have a single electron which is subject to the Coulomb field of the singly-charged, positive nucleus.   The potential energy of this system is given by

$$V = -e^2/r \qquad (2.69)$$

Hence the Schroedinger equation, which again is best expressed in spherical polar coordinates, becomes

$$\frac{1}{r^2}\frac{\partial}{\partial r}\left(r^2\frac{\partial\Psi}{\partial r}\right) + \frac{1}{r^2\sin\theta}\frac{\partial}{\partial\theta}\left(\sin\theta\frac{\partial\Psi}{\partial\theta}\right) + \frac{1}{r^2\sin^2\theta}\frac{\partial^2\Psi}{\partial\phi^2}$$
$$+ \frac{2\mu}{\hbar^2}\left(E + \frac{e^2}{r}\right)\Psi = 0 \quad (2.70)$$

Because of the large difference in mass between the electron and the nucleus of the atom, the reduced mass μ is very similar to the electronic mass m.

Again we can separate the variables in the Schroedinger equation by writing

$$\Psi = R(r) \cdot Y(\theta, \phi) \tag{2.71}$$

and this yields

$$\frac{1}{R}\frac{\partial}{\partial r}\left(r^2\frac{\partial R}{\partial r}\right) + \frac{2\mu r^2}{\hbar^2}\left(E + \frac{e^2}{r}\right)$$

$$= \frac{-1}{Y\sin\theta}\cdot\frac{\partial}{\partial\theta}\left(\sin\theta\frac{\partial Y}{\partial\theta}\right) - \frac{1}{Y\sin^2\theta}\frac{\partial^2 Y}{\partial\phi^2} \tag{2.72}$$

Now if we allow r to vary while keeping θ and ϕ constant, we see that the right-hand side of this equation is equal to a constant, which we shall call $l(l + 1)$. Hence we obtain two equations

$$\frac{1}{\sin\theta}\frac{\partial}{\partial\theta}\left(\sin\theta\frac{\partial Y}{\partial\theta}\right) + \frac{1}{\sin^2\theta}\frac{\partial^2 Y}{\partial\phi^2} + l(l+1)Y = 0 \tag{2.73}$$

and $$\frac{1}{r^2}\frac{d}{dr}\left(r^2\frac{dR}{dr}\right) + \left[\frac{2\mu}{\hbar^2}\left(E + \frac{e^2}{r}\right) - \frac{l(l+1)}{r^2}\right]R = 0 \tag{2.74}$$

Equation (2.73) is seen to be identical with (2.30), so that our eigenfunctions are of the form

$$\Psi = R(r) \cdot Y_{lm}(\theta, \phi) \tag{2.75}$$

where the Y_{lm} are normalized spherical harmonics.

In fact it happens whenever the potential energy V of a system can be expressed as a function of r only, that the *angular* part of the eigenfunctions is given by the spherical harmonics Y_{lm}. The quantum numbers l and m here refer effectively to the orbital angular momentum of the electron, since the nucleus is practically stationary by comparison.

To facilitate the process of solving for $R(r)$ in equation (2.74), we make the substitutions

$$E = -\mu e^4/2n^2\hbar^2 \tag{2.76}$$

and $$r = n\hbar^2 x/2\mu e^2 \tag{2.77}$$

where n is a parameter and x is a new variable. This procedure is not necessary to the argument but it greatly simplifies the algebra. Equation (2.74) now becomes

$$\frac{d^2R}{dx^2} + \frac{2}{x}\frac{dR}{dx} + \frac{n}{x} - \frac{l(l+1)}{x^2} - \frac{1}{4}R = 0 \tag{2.78}$$

Now if, after a process of trial and error, we decide to seek a solution in the form

$$R = u(x) \cdot x^l \cdot e^{-x/2} \qquad (2.79)$$

we find that $u(x)$ must satisfy the equation

$$x\frac{d^2u}{dx^2} + (2l + 2 - x) \cdot \frac{du}{dx} + (n - l - 1)u = 0 \qquad (2.80)$$

which happens to be a standard mathematical form.

The solutions of equation (2.80) are known as Associated Laguerre Polynomials, $L_\alpha^\beta(x)$, where $\beta = 2l + 1$ and $\alpha = n + l$. To see how these polynomials arise, we consider the equation

$$x\frac{dy}{dx} + (x - \alpha)y = 0 \qquad (2.81)$$

which is satisfied by

$$y = x^\alpha e^{-x} \qquad (2.82)$$

If we differentiate equation (2.81) $\alpha + 1$ times (so that α is now required to be an integer), we obtain

$$x\frac{d^2z}{dx^2} + (x + 1)\frac{dz}{dx} + (\alpha + 1)z = 0 \qquad (2.83)$$

where
$$z = \frac{d^\alpha y}{dx^\alpha} = \frac{d^\alpha}{dx^\alpha}(x^\alpha e^{-x}) = e^{-x} \cdot L_\alpha(x) \qquad (2.84)$$

Equation (2.84) defines the Laguerre Polynomial $L_\alpha(x)$ of degree α. The differential equation for these polynomials is, therefore,

$$\frac{x \cdot d^2L_\alpha}{dx^2} + (1 - x)\frac{dL_\alpha}{dx} + \alpha L_\alpha = 0 \qquad (2.85)$$

When (2.85) is differentiated β times (so β is also necessarily an integer) the result is

$$x\frac{d^2L_\alpha^\beta}{dx^2} + (\beta + 1 - x)\frac{dL_\alpha^\beta}{dx} + (\alpha - \beta)L_\alpha^\beta = 0 \qquad (2.86)$$

Equation (2.86) is the differential equation for the Associated Laguerre Polynomial of degree $(\alpha - \beta)$, and it is seen to be identical with (2.80). General solutions of equations (2.85) and (2.86) exist in the form of infinite series with limited domains of convergence, but as usual we require convergence in the whole range of x, and so we are interested only in the solutions which reduce to polynomials. These occur when α and β are integers. This means that n and l are both required to be integers, and one consequence of this is that equation (2.76) now defines a series of quantized energy levels, which in fact are identical with those that are given by the Bohr theory. We observe that the Associated Laguerre Polynomial reduces to zero if β is greater than α.

Hence the *radial* part of the wave function for a hydrogen atom is

$$R = N \cdot x^l \cdot e^{-x/2} \cdot L_{n+l}^{2l+1}(x) \tag{2.87}$$

where
$$n \geqslant l + 1.$$

On returning to our original variable, r, it is useful to introduce the first Bohr radius a_0, where $x = 2r/na_0$.

The normalizing factor N can be shown (footnote 4) to be

$$\sqrt{\left(\frac{2}{na_0}\right)^3 \frac{(n-l-1)!}{2n[(n+l)!]^3}} \tag{2.88}$$

so that our complete eigenfunctions $\Psi_{n,l,m}$ for the hydrogen atom are of the form

$$\sqrt{\left(\frac{2}{na_0}\right)^3 \frac{(n-l-1)!}{2n \cdot [(n+l)!]^3}} \left(\frac{2r}{na_0}\right)^l \cdot e^{-r/na_0} \cdot L_{n+l}^{2l+1}\left(\frac{2r}{na_0}\right) \cdot Y_{lm}(\theta, \phi) \tag{2.89}$$

It is assumed that the reader is familiar with the qualitative features of these wave functions, such as the form of the radial and angular distributions of Ψ and Ψ^2, presence of nodes, degeneracies of $s, p, d, \cdots$ orbitals, and so on. In the case of hydrogen we have an accidental degeneracy, in that levels with a given value of n have almost precisely the same energy, regardless of the value of l. This is not true for atoms with more than one electron. The wave functions (2.89) are also valid for hydrogen-like atoms (He^+, Li^{++}, etc.) if a_0 is replaced everywhere by a_0/Z, where Ze is the nuclear charge. Finally it should be mentioned that in addition to the discrete spectrum of negative energy eigenvalues which are given by (2.76), there also exists a continuous spectrum of positive eigenvalues, corresponding to the situation where the electron is not bound to the nucleus of the atom. The continuous spectrum is not of chemical interest and will not be considered here.

EXERCISES

2.1 Consider a potential barrier of the form shown below.

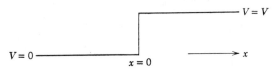

(i) Verify that the general solutions of the Schroedinger equation for the two regions are

$$\Psi_1 = Ae^{ik_1x} + Be^{-ik_1x} \qquad x < 0$$
$$\Psi_2 = Ce^{ik_2x} + De^{-ik_2x} \qquad x > 0$$

where $k_1\hbar = \sqrt{2mE}$, $k_2\hbar$ equals $\sqrt{2m(E - V)}$, and the positive sign in the exponential corresponds to the case of a particle moving from left to right (particle with positive momentum).

(ii) Calculate the probability that such a particle will be reflected at the barrier. [It must be assumed that both Ψ and $d\Psi/dx$ are continuous at $x = 0$. Start with the assumption that the particle moves only from left to right on the right of the barrier, and calculate $|\Psi|^2$ in both regions.]

2.2 Using equations (2.26), (2.27), and (2.28) work out the form of ∇ and ∇^2 in spherical polar coordinates, where

$$\nabla = \frac{\partial}{\partial x} + \frac{\partial}{\partial y} + \frac{\partial}{\partial z} \quad \text{and} \quad \nabla^2 = \nabla \cdot \nabla = \left(\frac{\partial}{\partial x}\right)^2 + \left(\frac{\partial}{\partial y}\right)^2 + \left(\frac{\partial}{\partial z}\right)^2$$

2.3 Prove $P_l^m(x) = 0$ for $m > l$ and $L_\alpha^\beta(x) = 0$ for $\beta > \alpha$.

2.4 Write out the first four Legendre, Hermite, and Laguerre Polynomials from the formulas (2.46), (2.64), and (2.84). Form in addition five or six of the corresponding Associated Legendre and Laguerre Polynomials.

2.5 (a) Show that $f_{mn}(t) = f_{mn}e^{i\omega_{mn}t}$ (where $f_{mn}(t)$ is a matrix element for time dependent wave functions, f_{mn} is the corresponding matrix element for stationary state wave functions, and $E_m - E_n = \hbar\omega_{mn}$). Hence show that $\dot{f}_{mn}(t) = i\omega_{mn}f_{mn}(t)$, where the dot signifies differentiation with respect to time

$$[\text{Assume } \Psi_m(q, t) = \phi_m(q) \cdot e^{-E_m it/\hbar}]$$

(b) Beginning from the equation of motion of the harmonic oscillator in the form $\ddot{x} + \omega^2 x = 0$, construct first the corresponding operator equation and then the corresponding matrix equation. (Simply convert $\ddot{x}$ and x into operators, let them operate on Ψ_n, then multiply by Ψ_m^* and integrate over $d\tau$.)

(c) Hence obtain the *selection rule*: that $x_{mn} = 0$ unless $\omega_{mn} = \pm\omega$. (Cf. section 7b.)

2.6 Compare the classical and quantum mechanical pictures of the probability distribution of the position of a particle undergoing simple harmonic motion, at low and high values of the vibrational quantum number.

2.7 Verify that the system of wave functions (2.89) gives rise to the usual shell structure of an atom.

3

Approximate Methods: Variation Principle and Perturbation Theory

....

3a INTRODUCTION

The variation principle and perturbation theory represent two alternative approaches to the problem of calculating approximate wave functions and energies of systems for which direct solution of the Schroedinger equation is difficult or impossible.   In this chapter we shall derive formulas which we will later apply to systems of chemical interest.   The applications themselves will not be discussed in this chapter, and the student may wish to omit the derivations at a first reading, merely noting the form and location of the results for later use.   Time-dependent perturbations are included here, although their application is of a different nature, being mainly concerned with the determination of the effects of varying external influences on systems whose stationary states are known.

3b VARIATION PRINCIPLE

The variation principle enables us either to select the best wave function from a number of different approximations or to determine an optimum value for a parameter in an approximate wave function. It depends on a theorem which may be stated simply in the form: "The best wave function is that which corresponds to the lowest energy."

The theorem is stated mathematically as

$$\int \Psi^* \hat{H} \Psi d\tau \geqslant E_0 \tag{3.1}$$

where Ψ is an approximate wave function and E_0 is the lowest eigenvalue of the accurate Hamiltonian $\hat{H}$. If Ψ is not already normalized, we must write instead

$$\frac{\int \Psi^* \hat{H} \Psi \, d\tau}{\int \Psi^* \Psi \, d\tau} \geqslant E_0 \tag{3.1a}$$

The theorem can be proved simply as follows:

If the approximate wave function Ψ is imagined to be expanded as a linear combination of the actual wave functions ϕ of the system, i.e., if we write

$$\Psi = a_0 \phi_0 + a_1 \phi_1 + a_2 \phi_2 + \cdots + a_r \phi_r \tag{3.2}$$

the left-hand side of the inequality (3.1) is recognized as the mean value of the energy of the system in the state Ψ (cf. equation 1.23). The mean value is given by

$$\bar{E} = |a_0|^2 \cdot E_0 + |a_1|^2 \cdot E_1 + \cdots + |a_r|^2 \cdot E_r \tag{3.3}$$

which is obviously greater than E_0 unless Ψ and ϕ_0 are identical, and the theorem is proved. It is an essential part of this proof that the actual wave functions ϕ form a complete orthonormal set, as discussed in section 1c. We note that as the approximate energy $\bar{E}$ decreases towards the value E_0 the coefficients $|a_1|^2$, $|a_2|^2$, $\cdots$ $|a_r|^2$ in equation (3.3) must tend to zero, while $|a_0|^2$ similarly tends to unity, because the effect of the terms in E_1, E_2, $\cdots E_r$ is always to make $\bar{E}$ larger than E_0. While this is happening to the energy, the approximate wave function Ψ must also be approaching closer and closer to ϕ_0, and this justifies our initial statement that the best wave function is that which corresponds to the lowest energy.

To make use of the theorem we generally form an approximate wave function which contains a number of arbitrary parameters. These parameters are then adjusted so as to minimize the integral (3.1), and the resulting wave function is the best available in the chosen form. It very often happens that a series of stationary values of the integral is obtained, corresponding to a series of energy levels for the system. A number of examples of calculations which depend on the variation principle will be found in Chapter 5.

3c PERTURBATION THEORY

In this section we consider only perturbations which are independent of time. Perturbation methods are most useful when dealing with a system which is not greatly different from one for which the eigenfunctions and eigenvalues have already been determined, i.e., when the difference between the new system and the old can be regarded as a small correction.

Let $\hat{H}_0$, $\Psi_n^{(0)}$ and $E_n^{(0)}$ refer to the unperturbed system, for which the solutions of the Schroedinger equation

$$\hat{H}_0\Psi_n^{(0)} = E_n^{(0)}\Psi_n^{(0)} \qquad (3.4)$$

are assumed to be known. For the perturbed system which we are studying we have

$$\hat{H}\Psi_n = E_n\Psi_n \qquad (3.5)$$

where

$$\hat{H} = \hat{H}_0 + \hat{H}' \qquad (3.6)$$

and $\hat{H}'$ represents a small perturbation. The equation to be solved is, therefore,

$$(\hat{H}_0 + \hat{H}')\Psi_n = E_n\Psi_n \qquad (3.7)$$

We first consider the case where there are no degenerate energy levels, that is, $E_n - E_k \neq 0$ for any value of n or k.

Let us expand Ψ_n in terms of the complete set of unperturbed eigenfunctions $\Psi_m^{(0)}$. We write

$$\Psi_n = \sum_m a_m\Psi_m^{(0)} \qquad (3.8)$$

Substitution of this expansion in equation (3.7) yields

$$\sum_m a_m\hat{H}_0\Psi_m^{(0)} + \sum_m a_m\hat{H}'\Psi_m^{(0)} = \sum_m a_mE_n\Psi_m^{(0)} \qquad (3.9)$$

that is,

$$\sum_m a_m(E_m^{(0)} + \hat{H}')\Psi_m^{(0)} = \sum_m a_mE_n\Psi_m^{(0)} \qquad (3.10)$$

Multiplying both sides of (3.10) by $\Psi_k^{(0)*}$ and integrating over $d\tau$, we obtain, since the $\Psi_n^{(0)}$ form an orthonormal set,

$$a_kE_k^{(0)} + \sum_m a_mH'_{km} = a_kE_n \qquad (3.11)$$

That is,
$$a_k(E_n - E_k^0) = \sum_m a_m H'_{km} \tag{3.12}$$

where
$$H'_{km} = \int \Psi_k^{(0)*} \cdot \hat{H}' \cdot \Psi_m^{(0)} \, d\tau$$

is a matrix element of the perturbation operator with respect to the unperturbed eigenfunctions.

Now we shall attempt to determine, by a successive approximation method, the energies and coefficients for the perturbed system, in the form

$$E_n = E_n^{(0)} + E_n^{(1)} + E_n^{(2)} + \cdots \tag{3.13}$$

$$a_k = a_k^{(0)} + a_k^{(1)} + a_k^{(2)} + \cdots \tag{3.14}$$

where successive terms in the series became progressively smaller and smaller. (Obviously this is only feasible if the perturbation is small.) Equation (3.14) implies that we are expanding Ψ_n as the series

$$\Psi_n = \Psi_n^{(0)} + \sum_{k \neq n} a_k^{(1)} \Psi_k^{(0)} + \sum_{k \neq n} a_k^{(2)} \Psi_k^{(0)} + \cdots \tag{3.15}$$

In this equation we have made use of the fact that Ψ_n tends to $\Psi_n^{(0)}$ as the perturbation tends to zero, so that $a_k^{(0)}$ is zero for $k \neq n$, while $a_n^{(0)} = 1$, and to avoid complications later we have excluded the terms involving $a_n^{(1)}$, $a_n^{(2)}$, etc., from the summations. All these coefficients will ultimately require some adjustment if Ψ_n is to be normalized.

One way of describing the nature of the expansion (3.15) is to say that the perturbation brings about a "mixing" of the original unperturbed states, and we shall see that the extent of mixing of two states is inversely proportional to the energy difference between them. It is, of course, inevitable that any change in one of the wave functions is equivalent to the mixing in of various proportions of the others, because the $\Psi_n^{(0)}$ comprise a *complete* orthonormal set for the system.

To find the first-order perturbation corrections we now substitute

$$E_n = E_n^{(0)} + E_n^{(1)} \tag{3.16}$$

$$a_k = a_k^{(0)} + a_k^{(1)} \tag{3.17}$$

in equation (3.12) and retain only first-order terms. With $k = n$ we obtain

$$E_n^{(1)} = \frac{\sum_m a_m H'_{nm}}{a_n}$$

$$= H'_{nn} \tag{3.18}$$

when we put $a_m^{(0)} = 0$ for $m \neq n$ and neglect second-order terms such as $a_m^{(1)} \cdot H'_{nm}$.

For $k \neq n$ we find similarly

$$a_k^{(1)} = \frac{H'_{kn}}{E_n^{(0)} - E_k^{(0)}} \tag{3.19}$$

The correction term $a_n^{(1)}$ remains undetermined because of the factor $E_n^{(0)} - E_k^{(0)}$ in the denominator of (3.19), and it is for this reason that we have absorbed it into the first term of the expansion (3.15).

The first-order correction to the eigenfunction $\Psi_n^{(0)}$ in the perturbed system is therefore given by

$$\Psi_n^{(1)} = \sum_{k \neq n} \frac{H'_{kn}}{E_n^{(0)} - E_k^{(0)}} \cdot \Psi_k^{(0)} \tag{3.20}$$

The condition for perturbation theory to be applicable is that $\Psi_n^{(1)}$ must be small, i.e., that H'_{kn} must be much less than the spacing between the unperturbed energy levels, $E_n^{(0)} - E_k^{(0)}$, for all values of k.

We can proceed to determine the second-order perturbation corrections by substituting

$$E_n = E_n^{(0)} + E_n^{(1)} + E_n^{(2)}$$

$$a_k = a_k^{(0)} + a_k^{(1)} + a_k^{(2)}$$

in equation (3.12) and obtain, for example,

$$E_n^{(2)} = \sum_{k \neq n} \frac{|H'_{kn}|^2}{E_n^{(0)} - E_k^{(0)}} \tag{3.21}$$

where we have used the fact that $\hat{H}'$ is Hermitian, that is, $H'_{kn} = H'_{nk}{}^*$. It is not usual to go far beyond the first-order perturbation corrections because it becomes easier to make use of the variation principle.

As a fairly simple example of a perturbed system we could consider a vibrating diatomic molecule. The unperturbed system in this case is a harmonic oscillator, for which the eigenfunctions and eigenvalues were calculated in Chapter 2. A molecule always deviates from the behavior that would be expected for a harmonic oscillator (the deviation is termed "anharmonicity") in a manner which becomes more pronounced as the vibrational quantum number increases. If the deviation is not too great, it can be accounted for by adding, as a perturbation, a term in x^3 to the harmonic oscillator potential $V = \frac{1}{2}kx^2$. The effect of anharmonicity on the spacing of the energy levels is illustrated qualitatively in Fig. 7. Spectroscopists commonly take account of still higher powers of the displacement x when dealing with actual molecules.[1]

The derivation just given breaks down if two or more of the unperturbed levels are degenerate ($E_n^{(0)} = E_k^{(0)}$), and in this case we first consider the effect of the perturbation on the degenerate levels alone.

Let $\Psi_1^{(0)}$, $\Psi_2^{(0)}$, $\cdots \Psi_s^{(0)}$ be a total of s degenerate eigenfunctions corresponding to a single eigenvalue $E_s^{(0)}$. An arbitrary wave function for this energy level is

$$\Psi = a_1^{(0)}\Psi_1^{(0)} + a_2^{(0)}\Psi_2^{(0)} + \cdots + a_s^{(0)}\Psi_s^{(0)} \tag{3.22}$$

If we write out equation (3.12) with $k = 1, 2, \cdots s$, and put

$$E_n = E_s^{(0)} + E^{(1)} \tag{3.23}$$

we obtain the set of s equations

$$a_k^{(0)} \cdot E^{(1)} = \sum_{m=1}^{s} a_m^{(0)} H'_{km} \qquad (k = 1, 2, \cdots s) \tag{3.24}$$

or

$$\sum_{m=1}^{s} (H'_{km} - E^{(1)} \cdot \delta_m^k) a_m^{(0)} = 0 \tag{3.25}$$

where by definition δ_m^k equals zero for $m \neq k$ and 1 for $m = k$.

Now (3.25) is a set of s linear, homogeneous equations in the s unknowns $a_m^{(0)}$ (the "secular equations"), and the condition that

[1] See, for example, G. Herzberg, "Spectra of Diatomic Molecules," *Molecular Spectra and Molecular Structure*, Vol. I, 2nd Edition, New York: Van Nostrand, 1950, p. 90.

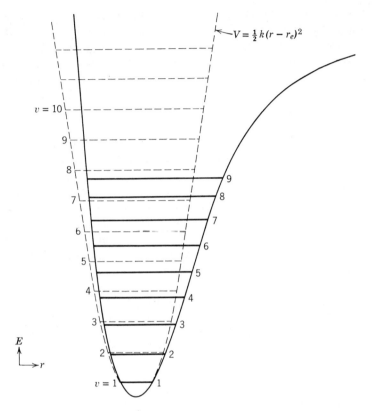

Fig. 7 The effect of a perturbation $\hat{H}' = g(r - r_e)^3 + \cdots$ on the energy levels of a harmonic oscillator (for which $V = \frac{1}{2}k[r - r_e]^2$). The solid curve is drawn approximately to scale for HCl, for which the energy levels are given by $E_v = hc\bar{\nu}_e(v + \frac{1}{2}) - hc\bar{\nu}_e x_e(v + \frac{1}{2})^2 + \cdots$ where $\bar{\nu}_e = 2989.74$ cm.$^{-1}$ and $\bar{\nu}_e x_e = 52.06$ cm.$^{-1}$.

there should be solutions of the equations other than $a_m^{(0)} = 0$ for all m is that the determinant of the coefficients should be zero, that is

$$|(H'_{km} - E^{(1)}\delta_m^k)| = 0 \qquad (3.26)$$

This is known as the "secular determinant." On expansion it becomes a polynomial of degree s in the first-order correction to the energy, $E^{(1)}$, and the s roots of this polynomial may be substituted in turn into (3.25) to enable the quantities $a_m^{(0)}$ to be evaluated.

Each value of $E^{(1)}$ gives rise to a different set of coefficients and thus to a different wave function of the perturbed system. In general the roots of the polynomial in $E^{(1)}$ are all different, and so the effect of the perturbation is to remove part or all of the degeneracy. The effects of including various perturbation terms on the calculated energy levels of a nitrogen atom are shown qualitatively in Fig. 13 (p. 105).

An equation of identical form to (3.26) occurs very frequently in Chapter 5 in connection with molecular orbital theory, and the type of results to be expected from a perturbation calculation with degenerate levels can be visualized fairly readily from the manner in which a set of atomic orbitals, having identical energies, may be combined in different proportions to form a set of molecular orbitals which in general are all of different energy.

In this book we shall not carry out any perturbation calculations for stationary states, but in several places (notably in Chapter 6) we shall have occasion to use the results of such calculations.

3d TIME-DEPENDENT PERTURBATIONS

Time-dependent perturbations are especially important in radiation theory, where we need to be able to calculate the effect of a varying external field on the rate at which transitions occur between the different energy levels of a system. The ideas which we shall introduce here, and consider in more detail in Chapter 7, are relevant to the whole field of spectroscopy.

Once more the Hamiltonian operator of the system is written

$$\hat{H} = \hat{H}_0 + \hat{H}' \tag{3.27}$$

but the perturbation $\hat{H}'$ is now a function of time.

In this situation it is necessary to make use of the Schroedinger equation including the time, i.e., for the unperturbed system we have

$$\hat{H}_0 \Psi^{(0)} = i\hbar \, \partial \Psi^{(0)}/\partial t \tag{3.28}$$

and for the perturbed system

$$(\hat{H}_0 + \hat{H}')\Psi = i\hbar \, \partial \Psi/\partial t \tag{3.29}$$

As usual we expand Ψ in terms of the unperturbed eigenfunctions $\Psi_n^{(0)}$, which are assumed to be of the form

$$\Psi_n^{(0)}(q, t) = \phi_n^{(0)}(q) \cdot e^{-iE_n t/\hbar} \tag{3.30}$$

That is, they correspond to stationary states of the system whose energies are E_n. The expansion is

$$\Psi(q, t) = \sum_n a_n(t) \cdot \Psi_n^{(0)}(q, t) \tag{3.31}$$

where we have expressly indicated that the coefficients in the expansion are functions of time.

When this expansion is substituted into (3.29) we obtain

$$\sum_n a_n \hat{H}_0 \Psi_n^{(0)} + \sum_n a_n \hat{H}' \Psi_n^{(0)} = i\hbar \sum_n \frac{da_n}{dt} \Psi_n^{(0)} + i\hbar \sum_n a_n \frac{\partial \Psi_n^{(0)}}{\partial t} \tag{3.32}$$

where all of the wave functions and coefficients are time dependent.

Equation (3.32) immediately reduces to

$$\sum_n a_n \cdot \hat{H}' \Psi_n^{(0)} = i\hbar \sum_n \frac{da_n}{dt} \cdot \Psi_n^{(0)} \tag{3.33}$$

If we multiply both sides of this equation by $\Psi_m^{(0)*}$ and integrate over $d\tau$ we obtain

$$\sum_n a_n \int \Psi_m^{(0)*} \cdot \hat{H}' \cdot \Psi_n^{(0)} \, d\tau = i\hbar \, da_m/dt \tag{3.34}$$

or $$\frac{da_m}{dt} = \frac{-i}{\hbar} \sum_n a_n \cdot H'_{mn}(t) \tag{3.35}$$

This equation gives, in effect, the rate at which the system undergoes transitions into the state $\Psi_m^{(0)}$ under the influence of the perturbation $\hat{H}'$. The actual probability of finding the system in this state at any given time is equal to $|a_m|^2$.

When m is allowed to take all possible values, the set of equations (3.35) can be solved to yield a series of first-order differential equations in the coefficients a_m. This provides a general approach to problems involving quantum effects which vary with time. The case where the system is initially in a definite state $\Psi_n^{(0)}(q, t)$ is especially simple, because then $a_n = 1$ and the other coefficients are zero, and (3.35) reduces to

$$\frac{da_m}{dt} = \frac{-i}{\hbar} \cdot H'_{mn}(t) \tag{3.36}$$

We note that the results so far have been expressed in terms of the matrix elements of $\hat{H}'$ with respect to the time-dependent wave functions $\Psi_m^{(0)}(q, t)$ (cf. Exercise 2.5). If H'_{mn} is a matrix element of $\hat{H}'$ with respect to the stationary state wave functions $\phi_m^{(0)}(q)$ then we have

$$H'_{mn}(t) = H'_{mn} \cdot e^{i(E_m - E_n)t/\hbar} \tag{3.37}$$

$$= H'_{mn} \cdot e^{i\omega_{mn}t} \tag{3.38}$$

where $\omega_{mn}(= 2\pi\nu_{mn})$ is the circular frequency of the radiation which is emitted or absorbed when a radiative transition occurs between the states m and n.

EXERCISES

3.1 Assuming a wave function $N \sin(kx)$ for a particle in a one-dimensional box of side a, use the variation principle to calculate the value of k. If desired, the normalizing factor may be taken in advance to be $(2/a)^{1/2}$.

3.2 Write out the secular determinant (3.26) in full for the case $s = 4$.

3.3 Work out the second-order perturbation correction to the eigenfunctions. (Answer:

$$\Psi_n^{(2)} = \sum_{k \neq n} \left[\sum_{m \neq n} \frac{H'_{km} \cdot H'_{mn}}{(E_n^{(0)} - E_k^{(0)})(E_n^{(0)} - E_m^{(0)})} - \frac{H'_{nn}H'_{kn}}{(E_n^{(0)} - E_k^{(0)})^2} \right] \Psi_k^{(0)})$$

4

The Use of Symmetry Properties and Group Theory

4a INTRODUCTION

In Chapter 2 we considered a few quantum mechanical systems for which the Schroedinger equation could be solved exactly. Such systems are very much in the minority, and for most cases of interest it is necessary to resort to the approximate methods outlined in Chapter 3. Indeed for a large number of interesting cases the difficulty of solving the Schroedinger equation is so great that we are driven to determining approximate eigenfunctions and eigenvalues for an approximate Hamiltonian.

It therefore comes as a relief to discover that by means of group theory we can exploit the symmetry properties of a quantum mechanical system of any degree of complexity in order to learn much of what we need to know about its eigenfunctions and eigenvalues, without the necessity of actually solving the Schroedinger equation. In addition to this, we can often greatly simplify the procedures for obtaining approximate solutions by introducing group theory at an early stage of the calculation. Furthermore, the methods of using group theory to attain these ends are not only simple in practice but are also mathematically exact. In the present account the emphasis will be on practical utility rather than mathematical rigor, and students who are interested in the latter aspect should consult the specialized books which are listed in Appendix I.

The introduction of group theory depends on the observation that the symmetry operations which leave a molecule unchanged form a group which is known as the *point group* of the molecule. The application to quantum mechanics arises from the fact that the Hamiltonian operator of a system must be invariant with respect to the symmetry operations which comprise the point group of the system. The implications of this statement will become clear later.

The main applications of group theory in quantum chemistry can be summarized as follows:

(i)   To label eigenfunctions in accordance with their symmetry properties.

(ii)   To predict the degeneracy of energy levels and the splitting of the degeneracy in fields of different symmetry.

(iii)   To form correct linear combinations of atomic orbitals in the L.C.A.O. method of molecular orbital theory.

(iv)   To determine numbers and symmetry types of normal vibrations of polyatomic molecules.

(v)   To determine selection rules for radiative transitions between energy levels of known symmetry.

Applications of group theory will form a large proportion of the material that is discussed in later chapters. For the present we shall concern ourselves with introducing the essential concepts of group theory and with justifying the small number of mathematical results that are actually required for these applications.

4b   THE MATHEMATICAL DEFINITION OF A GROUP

Any set of elements is said to form a group if it fulfills the following conditions:

1. The "product" of any two elements P and Q is also a member of the group. This implies that some meaningful way of combining the elements to form the product PQ exists and that

$$PQ = R \qquad (4.1)$$

where R is a member of the group.

2. The set contains the identity element E, where

$$RE = ER = R \qquad (4.2)$$

and R is any element of the group.

3. The associative law of multiplication holds with respect to the elements of the group, that is,

$$P(QR) = (PQ)R \qquad (4.3)$$

(The commutative law

$$PQ = QP \qquad (4.4)$$

does not necessarily hold.   If the commutative law holds, the group is said to be *Abelian*.)

4. Every element R in the group has an *inverse* R^{-1} which is also a member of the group, where

$$RR^{-1} = R^{-1}R = E \qquad (4.5)$$

A simple example of a group is provided by the collection of all positive and negative integers, including zero.   The "product" of A and B is defined as A plus B; the identity element is zero; and the inverse of A is $-A$.   This group is Abelian.

The number of elements in the group, infinity in this example, is called the *order* of the group.   Most of the groups with which we shall be concerned are of finite order, i.e., they contain a limited number of elements.

4c SYMMETRY OPERATIONS AND POINT GROUPS

The elements which go to make up point groups are the symmetry elements which correspond to symmetry operations about or through a fixed point.   This distinguishes point groups from space groups, in which symmetry elements involving translations, notably screw axes and glide planes, may also be included.   The method of combining two elements to form a product is to apply the corresponding symmetry operations in turn to some geometrical object. The product element is then one for which the corresponding symmetry operation has the same effect on the object as does the carrying out of the two component operations in turn.

If the carrying out of a particular symmetry operation leaves a molecule or other geometrical object unaltered, the object is said to possess the corresponding element of symmetry.   The relationship between symmetry operations and symmetry elements is illustrated in Table 4.1 for the symmetry elements which take part in point groups.

Table 4.1

Symbol	Symmetry Element	Symmetry Operation
E	Identity element	Leave molecule as it is.
C_n	n-fold axis of symmetry (proper axis)	Rotate about axis through $2\pi/n$ radians.
σ	Plane of symmetry	Reflect in the plane.
i	Center of symmetry	Invert through the center of symmetry.
S_n	n-fold alternating axis of symmetry (improper axis)	Rotate about axis through $2\pi/n$ radians, and reflect in a plane perpendicular to the axis.

For any symmetrical object the axis of symmetry with the largest value of n is called the principal axis. Planes perpendicular to this axis are labeled σ_h. Planes which contain the principal axis are labeled σ_v except that if they bisect the angles between n twofold axes perpendicular to the principal axis, they are labeled σ_d. (The letters h, v, and d are short for horizontal, vertical, and dihedral, respectively.) Some examples of the symmetry elements possessed by various molecules are given in Fig. 8.

Various combinations of symmetry elements to form point groups are possible. The most important types of point groups are listed in Table 4.2. The point groups are labeled by means of the Schoenflies symbols which are given in the first column of the table. An alternative system of symbols, in which for example C_{3v} becomes $3m$, is favored by crystallographers; however, the older Schoenflies notation is firmly established in the fields with which we shall be concerned.

The letters T and O designate tetrahedral and octahedral point groups, respectively. These groups have no principal axis and are often termed "special groups." The lists of symmetry elements given in Table 4.2 are far from exhaustive, for example, the presence of a C_6 axis implies the presence of coincident C_2 and C_3 axes, while the operation of a center of symmetry on the elements listed for O yields, as the components of O_h, altogether E, $8C_3$, $6C_2$, $3C_2'$, $6C_4$, i, $8S_6$, $6\sigma_d$, $3\sigma_h$, $6S_4$. (Here the last five classes of symmetry elements are obtained by allowing the center of symmetry to operate

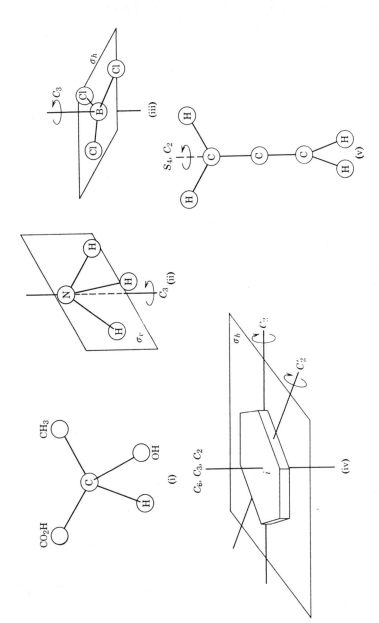

Fig. 8 Symmetry elements of molecules. (i) Lactic acid: element E only; (ii) ammonia: E, C_3, and $3\sigma_v$; (iii) boron tri-chloride: as NH_3, $+\sigma_h$; (iv) benzene: C_6, σ_v, σ_h, C_2, C_2', etc., i, etc.; (v) allene: S_4 and σ_v, C_2.

Table 4.2

Point Group	Essential Symmetry Elements
	Identity element, E, plus:
C_s	One symmetry plane
C_n	One n-fold axis of symmetry
S_n	One n-fold alternating axis (even n only, for n odd $\equiv C_{nh}$)
C_{nv}	Principal axis C_n plus n vertical planes σ_v
C_{nh}	Principal axis C_n plus a horizontal plane σ_h
D_n	Principal axis C_n plus nC_2 perpendicular to C_n
D_{nd}	D_n plus n dihedral planes σ_d
D_{nh}	D_n plus σ_h and $n\sigma_v$
T	3 mutually perpendicular C_2, $4C_3$, 4 independent C_3'.
T_d	T plus $6\sigma_d$ and $6S_4$, the $8C_3$ are now equivalent
O	$8C_3$, $6C_2$, $3C_2'$, $6C_4$
O_h	O plus a center of symmetry i

on the first five classes in order, that is, $i \times E = i$, $i \times C_3 = S_6$, etc.) The full complement of symmetry operations for any group may be found from the Character Tables that are given in Appendix II.

Certain point groups are sometimes given special symbols, for example $S_2 \equiv C_i$, $D_2 \equiv V$, $D_{2d} \equiv V_d$, $D_{2h} \equiv V_h$, $D_{3d} \equiv S_{6v}$, $D_{4d} \equiv S_{8v}$. The first of each of these pairs of symbols is the one that we shall use.

Linear molecules belong to the point groups $C_{\infty v}$ or $D_{\infty h}$, since the molecule is left unchanged by a rotation through any angle about the molecular axis. These are examples of *infinite* groups; all the other point groups with which we shall deal contain only a finite number of symmetry elements and are examples of *finite* groups.

At this stage it would be useful for the reader to work out the point groups of the molecules which are given in Exercise 4.1 and Fig. 8. The procedure for doing this is usually to decide first whether the molecule belongs to one of the special groups, T, T_d, O, or O_h. If this is not the case, the next step is to look for a principal axis, decide whether it is C_n or S_n, and then look for other symmetry elements in order to determine the complete point group.

4d REPRESENTATIONS OF GROUPS

Let us consider, as an example of a symmetrical object, the equilateral triangle of Fig. 9.   The point group here is D_{3h}, but we shall ignore the symmetry element σ_h in the plane of the paper and treat it as C_{3v}.   (C_{3v} is in fact a *sub-group* of D_{3h}.)   To distinguish the different rotations and reflections in this point group, we label them temporarily as follows: E, identity operation; A, reflection in yz plane; B, reflection in a similar plane through b; C, reflection in the similar plane through c; D, anticlockwise rotation through $2\pi/3$; F, clockwise rotation through $2\pi/3$.

Successive operation of any two symmetry elements is equivalent to a single application of one of the others.   Thus

$$EA(\Delta) = A(\Delta) \tag{4.6}$$

$$AB(\Delta) = F(\Delta) \tag{4.7}$$

(where B is applied before A) and so on.

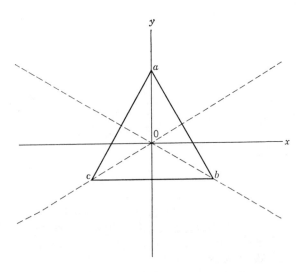

Fig. 9 Symmetry elements of an equilateral triangle.

Hence we can construct the "group multiplication table," which summarizes the results of combining all the symmetry operations in pairs. It reads:

		E	A	B	C	D	F	Applied second
Applied first	E	E	A	B	C	D	F	
	A	A	E	D	F	B	C	
	B	B	F	E	D	C	A	(4.8)
	C	C	D	F	E	A	B	
	D	D	C	A	B	F	E	
	F	F	B	C	A	E	D	

Any set of elements which are in $1:1$ correspondence with the elements E, A, B, C, D, F, and which obeys this group multiplication table also forms a group, which is said to be *isomorphic* with our original point group, C_{3v}.

The operation of a symmetry element is fundamentally the same thing as a transformation of coordinates, since it has the effect of changing the coordinates of a point from (x_1, y_1) to (x_2, y_2). Simple operations such as reflections and rotations correspond to linear transformations of coordinates, i.e., we can define them by means of equations of the form

$$x_2 = a_{11}x_1 + a_{12}y_1 \qquad (4.9)$$

$$y_2 = a_{21}x_1 + a_{22}y_1 \qquad (4.10)$$

Thus if we are considering a counterclockwise rotation through an angle θ (Fig. 10), we have

$$x_2 = x_1 \cos \theta - y_1 \sin \theta \qquad (4.11)$$

$$y_2 = x_1 \sin \theta + y_1 \cos \theta \qquad (4.12)$$

and for the reverse transformation

$$x_1 = x_2 \cos \theta + y_2 \sin \theta \qquad (4.13)$$

$$y_1 = -x_2 \sin \theta + y_2 \cos \theta \qquad (4.14)$$

as can be shown either by solving the equations or simply changing the sign of θ.

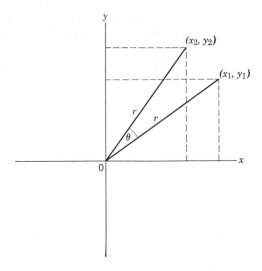

Fig. 10 Coordinate transformation during a rotation through an angle θ.

Substituting $\theta = 2\pi/3$ we find

$$\begin{pmatrix} x_2 \\ y_2 \end{pmatrix} = \begin{pmatrix} -1/2 & -\sqrt{3}/2 \\ +\sqrt{3}/2 & -1/2 \end{pmatrix} \cdot \begin{pmatrix} x_1 \\ y_1 \end{pmatrix} \qquad (4.15)$$

where we have used matrix notation to combine the two equations into one. The 2×2 matrix in (4.15) is called the matrix of the transformation, which in this case is the operation which was labeled D.

Similarly, we can obtain the whole array of transformation matrices:

$$\Gamma_3 \quad \underbrace{\begin{pmatrix} 1 & 0 \\ 0 & 1 \end{pmatrix}}_{E} \underbrace{\begin{pmatrix} -1 & 0 \\ 0 & +1 \end{pmatrix}}_{A} \underbrace{\begin{pmatrix} \frac{1}{2} & -\frac{\sqrt{3}}{2} \\ -\frac{\sqrt{3}}{2} & -\frac{1}{2} \end{pmatrix}}_{B} \underbrace{\begin{pmatrix} \frac{1}{2} & \frac{\sqrt{3}}{2} \\ \frac{\sqrt{3}}{2} & -\frac{1}{2} \end{pmatrix}}_{C} \underbrace{\begin{pmatrix} -\frac{1}{2} & -\frac{\sqrt{3}}{2} \\ \frac{\sqrt{3}}{2} & -\frac{1}{2} \end{pmatrix}}_{D} \underbrace{\begin{pmatrix} -\frac{1}{2} & \frac{\sqrt{3}}{2} \\ -\frac{\sqrt{3}}{2} & -\frac{1}{2} \end{pmatrix}}_{F}$$

$$(4.16)$$

For reasons which will appear later we label this array Γ_3. The matrix which corresponds to successive application of two symmetry operations is now obtained by applying the rule for matrix multiplication to the matrices of Γ_3. The matrices must obey the group multiplication table, and therefore they constitute a group which is isomorphic with C_{3v}. A set of matrices which obeys the group multiplication table in this way is called a *representation* of the group. Only square matrices can do this, and the number of rows or columns in the matrices is called the dimension of the representation. Thus the dimension of Γ_3 is two. The simplest representations of a group must be of unit dimension and for the point group C_{3v} two of these are possible, namely

	E	A	B	C	D	F	
Γ_1	1	1	1	1	1	1	(4.17)
Γ_2	1	−1	−1	−1	1	1	(4.18)

Matrices which are able to form representations of point groups are members of the class of *unitary* matrices. The elements of unitary matrices are required to fulfill the following conditions:

$$\sum_{j=1}^{n} A_{jk}^{*} \cdot A_{jk} = 1 \qquad \text{(sum down column)} \qquad (4.19)$$

$$\sum_{k=1}^{n} A_{jk}^{*} \cdot A_{jk} = 1 \qquad \text{(sum along row)} \qquad (4.20)$$

$$\sum_{j=1}^{n} A_{jk}^{*} \cdot A_{jl} = 0 \qquad \text{(different columns)} \qquad (4.21)$$

$$\sum_{k=1}^{n} A_{ik}^{*} \cdot A_{jk} = 0 \qquad \text{(different rows)} \qquad (4.22)$$

where A_{ij} is the matrix element in the ith row and jth column. In addition to these requirements it can be shown that the determinant of (A), which we write as $|A_{ij}|$, is such that $|A_{ij}|^{*} \cdot |A_{ij}| = 1$. If the matrix elements A_{ij} are all real, as is the case with the matrices of ordinary transformations (for example, the matrices in 4.16), then this result becomes

$$\text{Determinant of } (A) = |A_{ij}| = \pm 1 \qquad (4.23)$$

Unitary matrices have the useful property that

$$(A_{ij})^{-1} = (A_{ji}^{*}) \qquad (4.24)$$

That is, we can form the reciprocal of a unitary matrix by simply taking the complex conjugate and interchanging rows and columns. (This is known as forming the Hermitian conjugate. Cf. equations (4.11) to (4.14).) In the case where the matrix elements are all real ($A_{ij}^* = A_{ij}$) the foregoing conditions form the definition of an *orthogonal* matrix. We shall not trouble to distinguish between these two types, since any results which we derive for unitary matrices must also apply to orthogonal ones.

4e GROUP CHARACTERS AND REDUCIBLE REPRESENTATIONS

We can easily construct further sets of matrices which form representations of the group C_{3v}. For example, a four-dimensional representation is

$$\Gamma_4 =$$

$$
E \qquad
\begin{pmatrix}
1 & 0 & 0 & 0 \\
0 & 1 & 0 & 0 \\
0 & 0 & 1 & 0 \\
0 & 0 & 0 & 1
\end{pmatrix}
\qquad
A \qquad
\begin{pmatrix}
1 & 0 & 0 & 0 \\
0 & -1 & 0 & 0 \\
0 & 0 & -1 & 0 \\
0 & 0 & 0 & 1
\end{pmatrix}
\qquad
B \qquad
\begin{pmatrix}
1 & 0 & 0 & 0 \\
0 & -1 & 0 & 0 \\
0 & 0 & \tfrac{1}{2} & -\tfrac{\sqrt{3}}{2} \\
0 & 0 & -\tfrac{\sqrt{3}}{2} & -\tfrac{1}{2}
\end{pmatrix}
$$

$$
C \qquad
\begin{pmatrix}
1 & 0 & 0 & 0 \\
0 & -1 & 0 & 0 \\
0 & 0 & \tfrac{1}{2} & \tfrac{\sqrt{3}}{2} \\
0 & 0 & \tfrac{\sqrt{3}}{2} & -\tfrac{1}{2}
\end{pmatrix}
\qquad
D \qquad
\begin{pmatrix}
1 & 0 & 0 & 0 \\
0 & 1 & 0 & 0 \\
0 & 0 & -\tfrac{1}{2} & -\tfrac{\sqrt{3}}{2} \\
0 & 0 & \tfrac{\sqrt{3}}{2} & -\tfrac{1}{2}
\end{pmatrix}
\qquad
F \qquad
\begin{pmatrix}
1 & 0 & 0 & 0 \\
0 & 1 & 0 & 0 \\
0 & 0 & -\tfrac{1}{2} & \tfrac{\sqrt{3}}{2} \\
0 & 0 & -\tfrac{\sqrt{3}}{2} & -\tfrac{1}{2}
\end{pmatrix}
$$

$$(4.25)$$

It is fairly obvious that the preceding representation is made up from the three already considered, since the matrices are of the form

$$\begin{pmatrix} \Gamma_1 & 0 & 0 & 0 \\ 0 & \Gamma_2 & 0 & 0 \\ 0 & 0 & \Gamma_3 \\ 0 & 0 \end{pmatrix} \qquad (4.26)$$

and this representation is said to be reducible to $\Gamma_1 + \Gamma_2 + \Gamma_3$. If a representation cannot be reduced to a sum of representations of lower dimension, it is said to be irreducible. Γ_1, Γ_2, and Γ_3 are irreducible representations of the point group C_{3v}.

Most of our applications of group theory will involve finding the irreducible representations which are contained in various reducible representations, and we need to be able to do this even when the reducible representation is not in the diagonal form (4.26).

A unitary matrix may in principle be changed into the diagonal form by carrying out a series of similarity transformations. If A is the original matrix and X is any other unitary matrix of the same dimension, then a similarity transformation is carried out by forming the product

$$B = X^{-1}AX \qquad (4.27)$$

With large matrices this can be a tedious process and most computer centers possess library programs for carrying out matrix diagonalization. If, however, it is merely required to determine the components of the reducible representation, the diagonalization procedure is not necessary, since the information can be obtained directly from the *characters* of the reducible representation.

The character $\chi_i(R)$ of element R in the representation i is defined as the sum of the diagonal elements of the matrix of the symmetry element R in the ith representation, that is,

$$\chi_i(R) = R_{11} + R_{22} + \cdots + R_{hh} \qquad (\text{in } \Gamma_i) \qquad (4.28)$$

(In matrix algebra this quantity is called the *trace* of the matrix.)

In the example (4.25) it is easy to see that

$$\chi_4(R) = \chi_1(R) + \chi_2(R) + \chi_3(R) \qquad (4.29)$$

where R is any element of C_{3v}, but it is not obvious whether this relation holds if the matrices are not in the diagonal form. We shall now prove that the character of a representation is unchanged by a similarity transformation, from which it follows that the relation exemplified by (4.29) is true in all cases.

We have

$$\chi(P) = \sum_j P_{jj} \tag{4.30}$$

and

$$Q = X^{-1}PX = X^{-1} \cdot (PX) \tag{4.31}$$

where

$$\chi(Q) = \sum_i Q_{ii} \tag{4.32}$$

By the matrix multiplication rule the elements of the matrix (PX) are given by

$$(PX)_{mn} = \sum_k P_{mk} X_{kn} \tag{4.33}$$

Now since X is unitary

$$(X^{-1})_{rs} = X_{sr}^* \tag{4.34}$$

so

$$Q_{rs} = \sum_m (X^{-1})_{rm} (PX)_{ms} \tag{4.35}$$

$$= \sum_m X_{mr}^* \sum_k P_{mk} X_{ks} \tag{4.36}$$

$$= \sum_m \sum_k X_{mr}^* P_{mk} X_{ks} \tag{4.37}$$

Hence

$$\chi(Q) = \sum_i Q_{ii} = \sum_i \sum_m \sum_k X_{mi}^* X_{ki} P_{mk} \tag{4.38}$$

Now because X is unitary all terms in the sum add up to zero except those for which $m = k$, and since $\sum_k X_{ki}^* X_{ki} = 1$, we obtain

$$\chi(Q) = \sum_i Q_{ii} = \sum_k P_{kk} = \chi(P) \tag{4.39}$$

and the theorem is proved.

Tables of characters of irreducible representations of point groups which are relevant to quantum chemistry are given in Appendix II. The character table for point group C_{3v} is listed as

C_{3v}			E	$2C_3$	$3\sigma_v$
$(x^2 + y^2), z^2$	z	A_1	1	1	1
	R_z	A_2	1	1	-1
$(x^2 - y^2, xy)$	(x, y)	$\left.\right\} E$	2	-1	0
(xz, yz)	(R_x, R_y)				

$$(4.40)$$

The entries in the first two columns of the table are important in connection with spectroscopy and will be discussed in Chapter 7. The third column contains the conventional labels for the irreducible representations: our Γ_1 becomes A_1, Γ_2 becomes A_2, and Γ_3 becomes E. The letters A and B are reserved for one dimensional representations, A_1 always being the totally symmetric representation for which all matrices are $(+1)$. The letter B is used for representations which are antisymmetrical with respect to rotations about the principal axis, i.e., if there were a one-dimensional representation of the group C_{3v} for which the character of C_3 was -1, this representation would be labeled B. In fact no such representation exists. Two-dimensional representations are designated by the letter E and three-dimensional representations by T, or sometimes F. We note that the character of the identity element E is always equal to the dimension of the representation.

The two rotations C_3 and three reflections σ_v are each given a single column in the character table. Each of these sets of elements constitutes a single *class*, and it is usually possible to separate the elements into classes by inspection. The definition of a class is that two elements A and B belong to the same class if

$$B = X^{-1}AX \qquad (4.41)$$

where X is any other member of the group. Since (4.41) amounts to a similarity transformation it is apparent that all members of a class have the same character, and hence we can speak of the character of a class, as well as of a single operation. The identity element E always forms a class by itself.

Three theorems which are relevant to our applications of group theory will now be stated without proof. Proofs of these theorems may be found in several of the references which are given in Appendix I. The most important of the three from our point of view is:

1. *The number of irreducible representations of a group is equal to the number of classes.* This means that it is always possible to work out the number of times an irreducible representation occurs in a reducible representation from the characters alone, since for every class K there will be an equation of the type

$$\chi(K) = a_1\chi_1(K) + a_2\chi_2(K) + \cdots + a_n\chi_n(K) \qquad (4.42)$$

where a_m is the number of times the mth irreducible representation occurs in the reducible representation. Here $\chi(K)$ is the character of class K in the reducible representation, and $\chi_m(K)$ is the character of K in the mth irreducible representation. The number of equations is equal to the number of classes, which in turn is equal to the number of unknowns, a_m.

A useful expression of the general result is

$$a_m = 1/h \sum_R \chi(R) \cdot \chi_m(R) \qquad (4.43)$$

Here h is the order of the group, equal to 6 for C_{3v}, and the summation is carried out over all elements R of the group and not merely over all classes, K. We shall have considerable use for equation (4.43) in subsequent chapters.

As a simple illustration of the use of equation (4.43), suppose that by some means we have obtained a representation Γ_5 of C_{3v} for which the characters are as follows:

	E	$2C_3$	$3\sigma_v$
Γ_5	5	2	-1

Equation (4.43) now reads

$$a(A_1) = \tfrac{1}{6}(1 \times 5 \times 1 + 2 \times 2 \times 1 + 3 \times -1 \times 1)$$
$$= 1$$

and $$a(A_2) = \tfrac{1}{6}(1 \times 5 \times 1 + 2 \times 2 \times 1 + 3 \times -1 \times -1)$$
$$= 2$$

and $$a(E) = \tfrac{1}{6}(1 \times 5 \times 2 + 2 \times 2 \times -1 + 0)$$
$$= 1$$

Hence, $\Gamma_5 = A_1 + 2A_2 + E$

That is, the reducible representation contains A_1 once, A_2 twice, and

E once. Note that when we were forming the terms $\chi(R) \cdot \chi_m(R)$ we multiplied each term by the number of elements of type *R* which the group contains: that is, by 2 for $R = C_3$ and by 3 for $R = \sigma_v$.

The other important theorems are:

2. *The characters of the irreducible representations of a group behave as orthogonal vectors, that is,*

$$\sum_R \chi_i(R)\chi_j(R) = 0 \qquad i \neq j \tag{4.44}$$

and

$$\sum_R [\chi_i(R)]^2 = h \tag{4.45}$$

where *h* is the order of the group.

3. *The sum of the squares of the dimensions of the irreducible representations of a group is equal to the order of the group.*

$$d_1^2 + d_2^2 + \cdots + d_n^2 = h \tag{4.46}$$

or

$$\sum_i [\chi_i(E)]^2 = h \tag{4.47}$$

These last two theorems are sometimes useful in checking calculations.

4f RELATIONSHIP OF GROUP THEORY TO QUANTUM MECHANICS

The operation of a symmetry element on a molecule leaves the configuration of the molecule unchanged, hence it cannot alter either the energy or the Hamiltonian. The symmetry operation may, however, affect the wave functions.

If $\hat{R}$ is some symmetry operator, we have

$$\hat{R} \cdot \hat{H} \cdot \Psi_n = \hat{R}E_n \cdot \Psi_n \tag{4.48}$$

and, because $\hat{H}$ and *E* are unaffected by $\hat{R}$,

$$\hat{H}(\hat{R}\Psi_n) = E_n(\hat{R}\Psi_n) \tag{4.49}$$

where we have allowed $\hat{R}$ to operate on both the molecule and the wave function Ψ_n.

Hence if the energy level is nondegenerate, we can only have

$$\hat{R}\Psi_n = \pm\,\Psi_n \tag{4.50}$$

If the energy level is degenerate, the energy remains unaltered if the symmetry operation transforms Ψ into a linear combination of the degenerate eigenfunctions. Therefore we may have in general

$$\hat{R}\Psi_k = \sum_m R_{mk}\Psi_m \tag{4.51}$$

where the summation is carried out over all of the degenerate eigenfunctions.

If we choose the Ψ_m to be orthogonal to one another, then normalization requires

$$\sum_m |R_{mk}|^2 = 1 \tag{4.52}$$

Similarly, if $\hat{S}$ is another symmetry operation,

$$\hat{S}\Psi_m = \sum_q S_{qm}\Psi_q \tag{4.53}$$

Thus
$$(\hat{S}\hat{R})\Psi_k = \sum_q \sum_m S_{qm}R_{mk}\Psi_q \tag{4.54}$$

$$= \sum_q T_{qk}\Psi_q \tag{4.55}$$

where $(\hat{S}\hat{R}) = \hat{T}$ and

$$T_{qk} = \sum_m S_{qm}R_{mk} \tag{4.56}$$

We observe that the R_{mk}, etc., obey the matrix multiplication rule and it follows, from the normalization condition and the fact that the original Ψ_m are orthogonal to one another, that these matrices are unitary. Since the matrices must also obey the group multiplication table they form a representation of the symmetry group of the molecule, and the eigenfunctions Ψ_m are said to form a *basis* for this representation. The dimension of the representation is seen to be equal to the degeneracy of the energy level.

A representation which is generated in this way is always irreducible; otherwise it would be possible to divide the Ψ_m into two or more sets, each of which transformed only within itself under the action of the symmetry operations, and there would then be no reason for these sets to correspond to the same energy level. The only exception to this is the rare case of accidental degeneracy, when two different solutions of the Schroedinger equation chance to have practically identical eigenvalues.

To illustrate these ideas let us consider a molecule of point group C_{3v}, for example, ammonia. The eigenfunctions associated with a molecule of this symmetry, whether they are electronic, vibrational, or rotational can only be of three types. These types are:

(i) A_1: Totally symmetric, unchanged by the action of any of the symmetry operations of the molecule, nondegenerate.

(ii) A_2: Also nondegenerate, unchanged by the rotation C_3, changed in sign by the reflections σ_v.

(iii) E: Doubly degenerate, two wave functions which are transformed into linear combinations of one another by the action of the symmetry operations C_3 and σ_v.

It is apparent that the requirements of group theory act as a vigorous constraint on the types of eigenfunction which a symmetrical system can possess, and that these requirements can be of assistance by limiting the area of investigation when we are trying to discover an unknown eigenfunction.

4g   THE DIRECT PRODUCT

Suppose we have a product wave function of the form

$$\Psi_{AB} = \phi_A \cdot \phi_B \qquad (4.57)$$

where the symmetry types of the functions ϕ_A and ϕ_B are known. In general such a product wave function is a basis for a reducible representation, i.e., it can give rise to several different wave functions, each of which transforms according to one irreducible component of the reducible representation. We need to be able to reduce the representation of Ψ in order to determine these components, therefore we require an expression for the character of this reducible representation.

Let us consider the general case where ϕ_A and ϕ_B are respectively m- and n-fold degenerate. We have

$$\hat{R}\phi_{Ai} = \sum_{j=1}^{m} A_{ji}\phi_{Aj} \qquad (4.58)$$

and
$$\hat{R}\phi_{Bk} = \sum_{h=1}^{n} B_{hk}\phi_{Bh} \qquad (4.59)$$

where $\hat{R}$ is some symmetry operation.

The direct product wave function is

$$\Psi_{Ai, Bk} = \phi_{Ai} \cdot \phi_{Bk} \qquad (4.60)$$

so
$$\hat{R}\Psi_{Ai, Bk} = \hat{R}\phi_{Ai}\phi_{Bk} = \sum_j \sum_h A_{ji}B_{hk}\phi_{Aj}\phi_{Bh} \qquad (4.61)$$

$$= \sum_j \sum_h C_{jh, ik}\phi_{Aj}\phi_{Bh} \qquad (4.62)$$

where
$$C_{jh, ik} = A_{ji} \cdot B_{hk} \qquad (4.63)$$

The character of this matrix is found by summing the coefficients of the transformation (4.62) for the diagonal terms, i.e., for those where the product on the right-hand side is $\phi_{Ai}\phi_{Bk}$. This gives

$$\chi(C) = \sum_j \sum_h C_{jh, jh} \qquad (4.64)$$

$$= \sum_j \sum_h A_{jj} \cdot B_{hh}$$

That is
$$\chi(C) = \chi(A) \cdot \chi(B) \qquad (4.65)$$

Therefore we have proved that the character of a symmetry operation in a direct product representation is equal to the product of the characters of the symmetry element in the constituent representations. Equations (4.43) and (4.65) summarize most of the knowledge of group theory that will be needed for our later work on the applications of group theory to quantum chemistry.

EXERCISES

4.1 Specify the point groups to which the following molecules belong: H_2O, CO_2, $BeCl_2$, C_2H_4, BF_3, PH_3, SF_6, CH_4. (*Answers:* C_{2v}, $D_{\infty h}$, $D_{\infty h}$, D_{2h}, D_{3h}, C_{3v}, O_h, T_d.)

4.2 Write out in full the transformation equations corresponding to the matrices (4.16).

4.3 Use the matrix multiplication rule to show that the definition of a unitary matrix, $(A_{ij}^*) \cdot (A_{ji}) = A^{-1} \cdot A = E$, leads to the requirements of equations (4.19) to (4.22).

4.4 Prove that the matrix (R_{mk}) of equation (4.51) is unitary.

4.5 Reduce the representation $\Gamma(AB)$ of the point group C_{3v} where $\Gamma(A)$ and $\Gamma(B)$ are isomorphic with the representations A_2 and E, respectively.

4.6 When an additional element of symmetry is added to a point group it often happens that the number of irreducible representations is thereby doubled, each of the original representations having the opportunity to be symmetric or antisymmetric with respect to the new symmetry element. With reference to the character tables, consider the effect of adding: (i) σ_v to C_2; (ii) σ_h to C_2; (iii) σ_h to D_3; (iv) i to D_6; (v) i to O, (vi) i to $C_{\infty v}$.

5

Hückel Molecular Orbital Theory and Applications to Organic Chemistry

5a INTRODUCTION

In this chapter we shall consider the practical problem of how to use molecular orbital theory to calculate wave functions and energy levels for electrons in molecules. The theory is quite capable of dealing with entire molecules,[1] but we shall limit our attention to electrons in π orbitals, especially those which occur in conjugated and aromatic systems. Despite this limitation the results which we shall obtain will be relevant to a major part of the field of organic chemistry, with particular application to questions of molecular stability, aromaticity, chemical reactivity, and electronic spectra. The principles involved are relatively straightforward, and this chapter can probably be read with little reference to the remainder of the book, with the likely exception of Chapter 4. The applications of the theory are sufficiently varied to make it a valuable addition to the armory of any organic chemist, and several recent books have appeared with this idea in mind (see Appendix I).

In its simplest and most useful form the Hückel theory involves some fairly drastic approximations. It is possible to make the theory more sophisticated by introducing, for example, anti-symmetrized self-consistent-field molecular orbitals, but often the

[1] See, for example, R. Hoffmann, *J. Chem. Phys.*, **39**, 1397 (1963).

improved results are not very much more useful to the experimenter, while the labor of obtaining them is vastly increased. The simple theory, which provides answers in terms of one or two arbitrary parameters, is most useful as a theoretical framework for correlating the properties of members of a series of compounds. Since this, rather than energy levels correct to 1%, is generally what the chemist requires it seems permissible to leave the more precise calculations to specialist theoreticians. We shall note the sources of error in our calculations and consider how they might in principle be eliminated, but will otherwise be content to obtain our results by the simplest possible means.

5b THE HÜCKEL APPROXIMATION

We wish to find molecular orbitals Ψ in the form of linear combinations of atomic orbitals ϕ, that is,

$$\Psi = a_1\phi_1 + a_2\phi_2 + \cdots + a_n\phi_n \tag{5.1}$$

According to the variation principle the coefficients of the expansion, a_r, must be chosen so as to minimize the energy. This means that we need to determine the stationary values of the expression

$$E = \frac{\int \Psi^* \hat{H} \Psi \, d\tau}{\int \Psi^* \Psi \, d\tau} \tag{5.2}$$

where the factor in the denominator is necessary in case Ψ is not normalized. The quantity E will then always be greater than, or equal to, the true energy of the system, and in fact we shall behave from now on as if it were the true energy. It will in any case be the best value that can be obtained from the L.C.A.O. approximation (linear combination of atomic orbitals) with the available atomic orbitals.

We are usually able to arrange that the wave functions are entirely real, so that (5.2) becomes simply

$$E = \frac{\int \Psi \hat{H} \Psi \, d\tau}{\int \Psi^2 \, d\tau} \tag{5.3}$$

Now if we express Ψ in terms of the linear combination of atomic orbitals (5.1), we can locate the minima of (5.3) by putting

$$\partial E / \partial a_r = 0 \qquad \text{for all } r \tag{5.4}$$

We have

$$E = \frac{\int \sum_r (a_r\phi_r) \hat{H} \cdot \sum_r (a_r\phi_r) \, d\tau}{\int \left(\sum_r a_r\phi_r\right)^2 d\tau} \tag{5.5}$$

$$= \frac{\sum_r \sum_s a_r a_s \int \phi_r \hat{H}\phi_s \, d\tau}{\sum_r \sum_s a_r a_s \int \phi_r \phi_s \, d\tau}$$

$$= \sum_r \sum_s a_r a_s H_{rs} \Big/ \sum_r \sum_s a_r a_s S_{rs} \tag{5.6}$$

where H_{rs} is the matrix element $\int \phi_r \hat{H}\phi_s \, d\tau$ and S_{rs} is the *overlap integral* of ϕ_r and ϕ_s, given by $\int \phi_r \phi_s \, d\tau$.

When (5.6) is written out in full it reads

$$E(a_1a_1S_{11} + a_1a_2S_{12} + \cdots + a_1a_nS_{1n} + a_2a_1S_{21} + \cdots + a_na_nS_{nn})$$
$$= (a_1a_1H_{11} + a_1a_2H_{12} + \cdots + a_na_nH_{nn}) \tag{5.7}$$

Differentiating with respect to a_m yields, from (5.7),

$$\frac{\partial E}{\partial a_m}\left(\sum_r \sum_s a_r a_s S_{rs}\right) + E\left(\sum_r a_r S_{rm} + \sum_r a_r S_{mr}\right)$$
$$= \sum_r a_r H_{rm} + \sum_r a_r H_{mr} \tag{5.8}$$

Now $H_{rm} = H_{mr}$ (Hermitian condition)

and $S_{rm} = S_{mr}$ (for real wave functions)

and $\partial E/\partial a_m = 0$ (for E to be a minimum)

Hence we obtain the n equations ($m = 1, 2, \cdots n$)

$$\sum_{r=1}^{n} a_r(H_{rm} - ES_{rm}) = 0 \tag{5.9}$$

This is a system of n equations in the $(n - 1)$ unknowns,[2] a_r/a_1, and the condition that there should be solutions, other than $a_r = 0$

---

[2] Alternatively we can include the normalizing condition $\int \Psi^2 \, d\tau = 1$ and so obtain $(n + 1)$ equations in the n unknowns, a_r. The above is simpler.

for all r, is that the determinant of the coefficients should be zero, that is,

$$|H_{rm} - ES_{rm}| = 0 \qquad (5.10)$$

The determinant in (5.10) is known as the "secular determinant." On expansion it yields a polynomial of degree n in the energy E. The n roots of this polynomial are the stationary values of E and thus form a series of energy levels for the molecule. These values may be substituted in turn into the equations (5.9), after which these equations can be solved for the corresponding values of the coefficients a_r, provided the integrals H_{rs} and S_{rs} can be evaluated.

The approximations which were introduced by Hückel are intended to simplify the secular determinant so that the calculation of energy levels becomes relatively easy. The approximations are as follows:

(i)   The *Coulomb integrals* H_{rr} are put equal to a constant α for all carbon atoms in the molecule. If we were dealing with both σ and π orbitals, there would be a different α for each type of atomic orbital involved. Since, however, we are assuming the framework of σ bonds to be already fixed and are considering only the π electrons, we have only one sort of Coulomb integral.

(ii)   All *overlap integrals* S_{rs} are put equal to zero except where r equals s, when they are put equal to unity. This assumes, in effect, that the atomic orbitals form an orthonormal set. In fact, S_{rs} is about 0.25 for $2p_z$ orbitals of adjacent bonded carbon atoms, but allowing for this nonorthogonality increases the work of calculation without adding appreciably to the accuracy of the results (cf. Exercise 5.5).

(iii)   The *resonance integrals* H_{rs} are put equal to β when atoms r and s are bonded to each other and equal to zero otherwise.

The secular determinant now has the appearance

$$
\begin{vmatrix}
(\alpha - E) & \beta_{12} & \beta_{13} & \cdots & \beta_{1n} \\
\beta_{21} & (\alpha - E) & \beta_{23} & \cdots & \beta_{2n} \\
\cdot & \cdot & \cdot & \cdots & \cdot \\
\beta_{n1} & \beta_{n2} & \beta_{n3} & \cdots & (\alpha - E)
\end{vmatrix} = 0 \qquad (5.11)
$$

where β_{rs} is equal to β if atoms r and s are bonded to each other.

Both α and β are negative and are usually expressed either in kilocalories per mole or in electron volts (1 electron volt = 23.06 kcal./mole). The negative energy corresponds to the formation of a bound system. We note that α is the energy of an electron which is subject to the Hamiltonian of the whole molecule while being held in a $2p_z$ orbital on one of the constituent carbon atoms. In practice both α and β are always regarded as parameters to be determined experimentally, rather than as theoretical entities to be obtained by integration, and often it is not even necessary to evaluate them explicitly.

Once the energy levels have been determined by solving equation (5.11) the available electrons are fed into these levels one by one, starting with the lowest and allowing two electrons per orbital in accordance with the Pauli exclusion principle. If there are too few electrons to fill a set of degenerate molecular orbitals (i.e., orbitals which correspond to the same value of E owing to the occurrence of multiple roots of equation 5.11, then we make use of Hund's rule, according to which the electrons remain in separate orbitals, with parallel spins, for as long as possible during the building up (*aufbau*) process. The actual expressions for the molecular orbitals are found by solving the simplified form of equations (5.9) for the coefficients a_r. It is important to remember that when an energy level is degenerate, any linear combination of the degenerate orbitals is an acceptable wave function for this energy level of the molecule. In section 5c we shall work out some typical examples.

5c   EXAMPLES OF MOLECULAR ORBITAL CALCULATIONS

Ethylene

$$\underset{1}{\text{O}}\text{———}\underset{2}{\text{O}}$$

The σ-bond framework, based on carbon sp^2 hybrid orbitals, is assumed to be already established. For the π electrons we have

$$\Psi = a_1\phi_1 + a_2\phi_2 \qquad (5.12)$$

The secular determinant is, therefore,

$$\begin{vmatrix} \alpha - E & \beta \\ \beta & \alpha - E \end{vmatrix} = 0 \qquad (5.13)$$

It is customary to simplify the determinant further by substituting

$$x = (\alpha - E)/\beta \qquad (5.14)$$

so that

$$\begin{vmatrix} x & 1 \\ 1 & x \end{vmatrix} = 0 \qquad (5.15)$$

That is, $x = \pm 1$, $E = \alpha \mp \beta$. When these values are substituted in (5.9) the equations for the coefficients reduce to

$$a_1 \pm a_2 = 0 \qquad (5.16)$$

Hence we obtain the energy level diagram:

$$
\begin{array}{lll}
E = \alpha - \beta & \underline{\hspace{2cm}} & \Psi = \dfrac{1}{\sqrt{2}}(\phi_1 - \phi_2) \\[2mm]
E = \alpha & \text{-- -- -- -- --} & \qquad\qquad (5.17) \\[2mm]
E = \alpha + \beta & \underline{\hspace{0.8cm}} \uparrow\downarrow \underline{\hspace{0.8cm}} & \Psi = \dfrac{1}{\sqrt{2}}(\phi_1 + \phi_2)
\end{array}
$$

(The wave functions Ψ have been normalized.) We note that a bonding ($E = \alpha + \beta$) and an antibonding ($E = \alpha - \beta$) molecular orbital are disposed symmetrically about the energy level which would correspond to a nonbonding orbital ($E = \alpha$). There is a general rule (which an algebraically inclined student should have no difficulty in proving) that the sum of the roots of the secular determinant is zero. This will often provide a useful check on our calculations. In the ground state of ethylene both π electrons occupy the bonding molecular orbital and the total π-electron energy of the molecule is $2(\alpha + \beta)$. In this simple theory we make no allowance for any interelectron repulsion energy.

Butadiene

For simplicity we assume that the molecule is linear and that all carbon-carbon bonds are of equal length. The secular determinant is

$$\begin{vmatrix} x & 1 & 0 & 0 \\ 1 & x & 1 & 0 \\ 0 & 1 & x & 1 \\ 0 & 0 & 1 & x \end{vmatrix} = 0 \qquad (5.18)$$

where $x = (\alpha - E)/\beta$ as before.   This may be expanded as follows:

$$0 = x \begin{vmatrix} x & 1 & 0 \\ 1 & x & 1 \\ 0 & 1 & x \end{vmatrix} - \begin{vmatrix} 1 & 1 & 0 \\ 0 & x & 1 \\ 0 & 1 & x \end{vmatrix} = x^4 - 3x^2 + 1 \quad (5.19)$$

That is, $x^2 = \frac{1}{2}(3 \pm \sqrt{5})$

$$x = \pm \sqrt{\frac{3 \pm \sqrt{5}}{2}} = \pm 1.618, \ \pm 0.618 \quad (5.20)$$

The coefficients a_r may be calculated in a straightforward though tedious manner from

$$
\begin{aligned}
a_1 x + a_2 &= 0 \\
a_1 + a_2 x + a_3 &= 0 \\
a_2 + a_3 x + a_4 &= 0 \quad (5.21)\\
a_3 + a_4 x &= 0 \\
a_1^2 + a_2^2 + a_3^2 + a_4^2 &= 1
\end{aligned}
$$

The results, for each value of x, are

	x	a_1	a_2	a_3	a_4	
Ψ_1	-1.618	0.371	0.600	0.600	0.371	
Ψ_2	-0.618	0.600	0.371	-0.371	-0.600	(5.22)
Ψ_3	0.618	0.600	-0.371	-0.371	0.600	
Ψ_4	1.618	0.371	-0.600	0.600	-0.371	

Hence the energy level diagram for the molecule is

$$
\begin{aligned}
E &= \alpha - 1.618\beta &&\text{——————}&& \Psi_4 \\
E &= \alpha - 0.618\beta &&\text{——————}&& \Psi_3 \\
E &= \alpha &&\text{— — — —}&& \quad (5.23)\\
E &= \alpha + 0.618\beta &&\text{———}\uparrow\downarrow\text{———}&& \Psi_2 \\
E &= \alpha + 1.618\beta &&\text{———}\uparrow\downarrow\text{———}&& \Psi_1
\end{aligned}
$$

In the ground state the four π electrons occupy the bonding orbitals Ψ_1 and Ψ_2 as shown, so that the total π-electron energy of the molecule is $4\alpha + 4.472\beta$. The energy of two ethylene-type double bonds would be $4\alpha + 4\beta$, so that in the conjugated system there is an additional *delocalization energy* of 0.472β, which confers

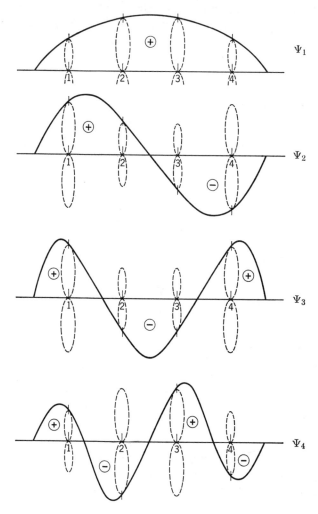

Fig. 11 Wave functions for butadiene.

additional stability on the molecule. This delocalization energy is obviously related to the "resonance energy" which occurs in valence bond theory.

The butadiene molecular orbitals may be visualized as in Fig. 11, where the component $2p_z$ wave functions have been sketched so that their size is proportional to their contribution to the molecular orbital.   Where a node (region of zero Ψ) occurs between two atoms the wave function is antibonding with respect to these atoms.   (For a nonbonding orbital the node occurs at the nonbonded atom.) The similarity of the wave functions of Fig. 11 to those of an electron in a one-dimensional box (section 2b) should be noted.

This would be a good time for the reader to attempt Exercise 5.1.

Butadiene: Use of Group Theory.   The results of Exercise 5.1 should convince the reader that the algebra involved in first solving the secular determinant to obtain x and then evaluating the coefficients of the atomic orbitals in the various molecular orbitals rapidly becomes disagreeably tedious and complex as the number of atoms in a molecule is increased.   In favorable cases we can greatly simplify these calculations by making use of group theory, as we shall now demonstrate for the case of butadiene (cf. Chapter 4).

The point group of butadiene is D_{2h}, however, we will neglect symmetry in the plane of the molecule and proceed as if it belonged to point group C_2.   The character table is

C_2	E	C_2
A	1	1
B	1	-1

$$(5.24)$$

and the correct molecular orbitals must transform as one of the representations A or B.   They are often called the correct "symmetry orbitals."

We take as the basis of a reducible representation the atomic orbitals ϕ_1, ϕ_2, ϕ_3, ϕ_4 and try to work out the character of each of the symmetry operations of the point group in this representation. If a particular symmetry operation leaves one of the basis functions unchanged, this function must contribute $+1$ to the character of the operation in the reducible representation.   If, however, the function is transformed into one of the other basis functions, the contribution to the character is zero.   Occasionally we may find that a symmetry operation has the effect of multiplying a basis function by -1, in which case the contribution to the character is also -1.   The character of the symmetry operation is found by summing the contributions from all of the basis functions.   These statements become self-evident if the symmetry operations are considered in terms of their transformation matrices. The procedure is best tabulated as in (5.25).

Basis Function ϕ	$\hat{E}\phi$	Contribution	$\hat{C}_2\phi$	Contribution
ϕ_1	ϕ_1	1	ϕ_4	0
ϕ_2	ϕ_2	1	ϕ_3	0
ϕ_3	ϕ_3	1	ϕ_2	0
ϕ_4	ϕ_4	1	ϕ_1	0
	Totals	4		0

$$(5.25)$$

Hence in the reducible representation $\chi(E) = 4$ and $\chi(C_2) = 0$. We reduce this representation by means of equation (4.43).

$$a_i = \frac{1}{h} \sum_R \chi_i(R) \cdot \chi(R)$$

Thus

$$a(A) = \tfrac{1}{2}(1 \times 4 + 1 \times 0) = 2$$

and

$$a(B) = \tfrac{1}{2}(1 \times 4 - 1 \times 0) = 2$$

so that the reducible representation is

$$\Gamma = 2A + 2B \qquad (5.26)$$

and the next step is to determine the appropriate symmetry orbitals.

A generally useful procedure at this stage is to form the sums

$$\Psi_i = \sum_R \chi_i(R) \cdot \hat{R}\phi_r \qquad (5.27)$$

where $\hat{R}\phi_r$ is the orbital which results from acting on ϕ_r with symmetry operation $\hat{R}$. These sums are linear combinations of atomic orbitals which, for one-dimensional representations at least, necessarily possess the correct symmetry properties. In favorable cases they are the actual symmetry orbitals of the molecule, otherwise they are simple linear combinations of the symmetry orbitals. For butadiene we obtain, after normalization,

$$\Psi_A = 1/\sqrt{2} \cdot (\phi_1 + \phi_4)$$
$$\Psi'_A = 1/\sqrt{2} \cdot (\phi_2 + \phi_3)$$
$$\Psi_B = 1/\sqrt{2} \cdot (\phi_1 - \phi_4)$$
$$\Psi'_B = 1/\sqrt{2} \cdot (\phi_2 - \phi_3)$$

$$(5.28)$$

Next we evaluate the matrix elements H'_{mn} in terms of α and β, where $H'_{11} = \int \Psi_A \hat{H} \Psi_A \, d\tau$, $H'_{12} = \int \Psi_A \hat{H} \Psi'_A \, d\tau$, etc. Thus, for example,

$$H'_{44} = \int \Psi'_B \hat{H} \Psi'_B \, d\tau = \tfrac{1}{2}(\int \phi_2 \hat{H} \phi_2 \, d\tau - \int \phi_2 \hat{H} \phi_3 \, d\tau$$
$$- \int \phi_3 \hat{H} \phi_2 \, d\tau + \int \phi_3 \hat{H} \phi_3 \, d\tau) = \alpha - \beta$$

and similarly

$$H'_{11} = H'_{33} = \alpha$$
$$H'_{12} = H'_{21} = H'_{34} = H'_{43} = \beta \tag{5.29}$$
$$H'_{13} = H'_{31} = H'_{14} = H'_{41} = H'_{23} = H'_{32} = H'_{24} = H'_{42} = 0$$
$$H'_{22} = \alpha + \beta$$

The secular determinant is now

$$\begin{vmatrix} x & 1 & 0 & 0 \\ 1 & x+1 & 0 & 0 \\ 0 & 0 & x & 1 \\ 0 & 0 & 1 & x-1 \end{vmatrix} = 0 \tag{5.30}$$

and by using group theory we have reduced the problem from that of solving a 4×4 determinant to that of solving two 2×2 determinants. If the orbitals (5.28) were the actual symmetry orbitals of the molecule, the problem would have reduced to solving four 1×1 determinants, i.e., the determinant would have been in the diagonal form. As before, we can now solve for x to find the energy levels and then use the two sets of linear equations corresponding to the determinant (5.30) to determine the coefficients of Ψ_A, Ψ_B, etc., in the symmetry orbitals. The reader is asked to do this in exercise 5.2.

Benzene

Without the aid of group theory we should be obliged to solve the equation

$$\begin{vmatrix} x & 1 & 0 & 0 & 0 & 1 \\ 1 & x & 1 & 0 & 0 & 0 \\ 0 & 1 & x & 1 & 0 & 0 \\ 0 & 0 & 1 & x & 1 & 0 \\ 0 & 0 & 0 & 1 & x & 1 \\ 1 & 0 & 0 & 0 & 1 & x \end{vmatrix} = 0 \tag{5.31}$$

which is a sixth degree polynomial in x.

The point group of benzene is D_{6h}; however, we shall again ignore symmetry in the plane of the molecule and proceed as if it belonged to point group C_6. As basis functions we use the atomic orbitals ϕ_1, ϕ_2, ϕ_3, ϕ_4, ϕ_5, and ϕ_6. The character table for C_6 is

C_6	E	C_6	C_3	C_2	$C_3^2 = C_3^{-1}$	$C_6^5 = C_6^{-1}$	
A	1	1	1	1	1	1	
B	1	-1	1	-1	1	-1	
E'	$\begin{cases} 1 \\ 1 \end{cases}$	ω ω^5	ω^2 ω^4	ω^3 ω^3	ω^4 ω^2	ω^5 ω	(5.32)
E''	$\begin{cases} 1 \\ 1 \end{cases}$	ω^2 ω^4	ω^4 ω^2	1 1	ω^2 ω^4	ω^4 ω^2	

Here $\omega = e^{2\pi i/6}$ is the complex sixth root of unity. We note that ω has the following interesting properties:

$$\omega \cdot \omega^* = 1 \qquad \text{that is, } \omega^* = \omega^{-1}$$

$$\omega^2 = -\omega^* = -\omega^{-1}$$

$$\omega^3 = -1, \omega^4 = -\omega, \omega^5 = -\omega^2 = \omega^{-1} \tag{5.33}$$

$$1 + \omega + \omega^2 + \omega^3 + \omega^4 + \omega^5 = 0$$

$$\omega + \omega^* = 1$$

It will be found to be an advantage to have two independent sets of characters for each of the two dimensional representations E' and

E''. We determine the characters of the reducible representation in the usual way, that is,

ϕ	$\hat{E}\phi$	$\hat{C}_6\phi$	$\hat{C}_3\phi$	$\hat{C}_2\phi$	$\hat{C}_3^2\phi$	$\hat{C}_6^5\phi$	
1	1	2	3	4	5	6	
2	2	3	4	5	6	1	
3	3	4	5	6	1	2	(5.34)
4	4	5	6	1	2	3	
5	5	6	1	2	3	4	
6	6	1	2	3	4	5	
$\chi(R) =$	6	0	0	0	0	0	

Hence, using (4.43), $a_i = 1/h \sum_R \chi_i(R) \cdot \chi(R)$

we find

$$\Gamma = A + B + E' + E'' \tag{5.35}$$

As before, we try to form symmetry orbitals using (5.27), $\Psi_i = \sum_R \chi_i(R) \cdot \hat{R}\phi_r$

This gives the following orbitals:

Type A $$\Psi_1 = \frac{1}{\sqrt{6}}(\phi_1 + \phi_2 + \phi_3 + \phi_4 + \phi_5 + \phi_6)$$

Type B $$\Psi_2 = \frac{1}{\sqrt{6}}(\phi_1 - \phi_2 + \phi_3 - \phi_4 + \phi_5 - \phi_6)$$

Type E'
$$\Psi_3 = \frac{1}{\sqrt{6}}(\phi_1 + \omega\phi_2 + \omega^2\phi_3 + \omega^3\phi_4 + \omega^4\phi_5 + \omega^5\phi_6)$$
$$\Psi_4 = \frac{1}{\sqrt{6}}(\phi_1 + \omega^5\phi_2 + \omega^4\phi_3 + \omega^3\phi_4 + \omega^2\phi_5 + \omega\phi_6)$$

Type E''
$$\Psi_5 = \frac{1}{\sqrt{6}}(\phi_1 + \omega^2\phi_2 + \omega^4\phi_3 + \phi_4 + \omega^2\phi_5 + \omega^4\phi_6)$$
$$\Psi_6 = \frac{1}{\sqrt{6}}(\phi_1 + \omega^4\phi_2 + \omega^2\phi_3 + \phi_4 + \omega^4\phi_5 + \omega^2\phi_6)$$

$$\tag{5.36}$$

where the normalizing factors have been determined with the aid of the relations (5.33) and using $\int \Psi^*\Psi \, d\tau = 1$.

We next determine the matrix elements H_{11}, H_{12}, etc., remembering that these will also involve complex conjugate wave functions, and so form the secular determinant. We find

$$\begin{vmatrix} \alpha + 2\beta - E & 0 & 0 & 0 & 0 & 0 \\ 0 & \alpha - 2\beta - E & 0 & 0 & 0 & 0 \\ 0 & 0 & \alpha + \beta - E & 0 & 0 & 0 \\ 0 & 0 & 0 & \alpha + \beta - E & 0 & 0 \\ 0 & 0 & 0 & 0 & \alpha - \beta - E & 0 \\ 0 & 0 & 0 & 0 & 0 & \alpha - \beta - E \end{vmatrix}$$

$$= 0 \qquad (5.37)$$

From the fact that the determinant is in the diagonal form we deduce that the orbitals (5.36) are the actual symmetry orbitals of the molecule.[3] The energy level scheme for benzene therefore has the appearance

$$
\begin{array}{lll}
B & \text{————————} & \alpha - 2\beta \\
E'' & \text{════════} & \alpha - \beta \\
& \text{— — — — — — —} & \alpha \qquad (5.38) \\
E' & \text{═══↑↓═══↑↓═══} & \alpha + \beta \\
A & \text{————↑↓————} & \alpha + 2\beta
\end{array}
$$

The total energy of the six π electrons is $6\alpha + 8\beta$, and by comparison with a system of three ethylenic double bonds we find the delocalization energy to be 2β, which is a significantly large value. The high stability and "aromaticity" of benzene are attributed to the large delocalization energy and to the presence of a "filled shell" of six electrons in the A and E' orbitals.[4] The degenerate wave functions in (5.36) are impossible to visualize (cf. Exercise 5.4) because of the complex coefficients. We can overcome this by

[3] The converse of this statement, that symmetry orbitals diagonalize the Hamiltonian and hence also the secular determinant, is not invariably true.

[4] The above labels refer to a system of point group C_6. In the correct point group D_{6h}, the A, E', E'', and B orbitals above transform according to the representations A_{2u}, E_{1g}, E_{2u}, and B_{1g}, respectively.

forming the linear combinations $\Psi_3 \pm \Psi_4$ and $\Psi_5 \pm \Psi_6$. This gives

Type E'
$$\begin{cases} \Psi'_3 = \dfrac{1}{\sqrt{12}}(2\phi_1 + \phi_2 - \phi_3 - 2\phi_4 - \phi_5 + \phi_6) \\[3mm] \Psi'_4 = \dfrac{1}{\sqrt{12}}(\phi_1 + 2\phi_2 + \phi_3 - \phi_4 - 2\phi_5 - \phi_6) \end{cases}$$

(5.39)

Type E''
$$\begin{cases} \Psi'_5 = \dfrac{1}{\sqrt{12}}(2\phi_1 - \phi_2 - \phi_3 + 2\phi_4 - \phi_5 - \phi_6) \\[3mm] \Psi'_6 = \dfrac{1}{\sqrt{12}}(\phi_1 - 2\phi_2 + \phi_3 + \phi_4 - 2\phi_5 + \phi_6) \end{cases}$$

Trimethylene Methane. In the previous example we were greatly assisted by the fact that two different sets of characters were listed for each of the two-dimensional representations, E' and E''. Other point groups for which this happens are C_3, C_4, C_5, C_6, C_{3h}, S_4, and T (for E but not for T representations), and it is usually best to treat molecules of higher symmetry as if they belonged to one of these point groups if degenerate energy levels are involved. It is always necessary to take the full symmetry of a molecule into consideration when dealing with spectra (cf. Chapter 7) but otherwise the assumption of a lower symmetry cannot invalidate the results. As an illustration of the difficulties which arise when this advice is not heeded we shall consider a simple example, namely trimethylene methane, which will be treated according to the point group C_{3v}.

The character table is

C_{3v}	E	$2C_3$	$3\sigma_v$
A_1	1	1	1
A_2	1	1	-1
E	2	-1	0

(5.40)

We first derive the characters of the reducible representation for which ϕ_1, ϕ_2, ϕ_3, and ϕ_4 provide a basis.

ϕ	$\hat{E}\phi$	$\hat{C}_3\phi$	$\hat{\sigma}_v\phi$
1	1	2	1
2	2	3	3
3	3	1	2
4	4	4	4
$\chi(R) =$ 4		1	2

$$(5.41)$$

Hence, using (4.43),

$$a(A_1) = \tfrac{1}{6}(4 \times 1 + 2 \times 1 + 2 \times 3) = 2$$
$$a(A_2) = \tfrac{1}{6}(4 \times 1 + 2 \times 1 - 2 \times 3) = 0 \qquad (5.42)$$
$$a(E) = \tfrac{1}{6}(4 \times 2 - 2 \times 1 + 2 \times 0) = 1$$

and so we have to find two nondegenerate orbitals of type A_1 and a pair of degenerate orbitals of type E. For the nondegenerate orbitals we use (5.27)

$$\Psi_i = \sum_R \chi_i(R) \cdot \hat{R}\phi_r$$

and obtain, after normalization

$$\Psi_1 = \frac{1}{\sqrt{3}} (\phi_1 + \phi_2 + \phi_3)$$
$$\Psi_2 = \phi_4 \qquad (5.43)$$

From these two functions we obtain the determinant

$$\begin{vmatrix} x & \sqrt{3} \\ \sqrt{3} & x \end{vmatrix} = 0 \qquad (5.44)$$

hence

$$E = \alpha + \sqrt{3} \cdot \beta \qquad (x = -\sqrt{3})$$

and

$$E = \alpha - \sqrt{3} \cdot \beta \qquad (x = +\sqrt{3})$$

$$(5.45)$$

are the nondegenerate energy levels. From the equations corresponding to the determinant (5.44),

$$a_1 x + a_2\sqrt{3} = 0$$
$$a_1\sqrt{3} + a_2 x = 0 \qquad (5.46)$$

we find that the actual symmetry orbitals are

$$\Psi(A_1) = \tfrac{1}{2}(\phi_1 + \phi_2 + \phi_3 + \phi_4)$$
$$\Psi'(A_1) = \tfrac{1}{2}(\phi_1 + \phi_2 + \phi_3 - \phi_4)$$

(5.47)

The first of these is the bonding orbital for which $E = \alpha + \sqrt{3} \cdot \beta$, and the second is the antibonding orbital for which $E = \alpha - \sqrt{3} \cdot \beta$.

If we apply equation (5.27) to the E-type representation, we obtain the *four* functions,

$$\Psi_3 = 2\phi_1 - \phi_2$$
$$\Psi_4 = 2\phi_2 - \phi_3$$
$$\Psi_5 = 2\phi_3 - \phi_1$$
$$\Psi_6 = \phi_4$$

(5.48)

and we are no further ahead. One way out of this difficulty is to expand the original secular determinant

$$\begin{vmatrix} x & 0 & 0 & 1 \\ 0 & x & 0 & 1 \\ 0 & 0 & x & 1 \\ 1 & 1 & 1 & x \end{vmatrix} = 0$$

(5.49)

and then remove the factor $(x^2 - 3)$ from the resulting polynomial. An alternative method, which requires some intuition, but which may save time with more complicated molecules, depends on the observation that the character of σ_v in the E representation is zero. This means that the required symmetry orbitals must be respectively symmetric and antisymmetric with respect to the operation of reflection in one of the σ_v planes of the molecule. Therefore we should try to form two independent linear combinations of the basis functions which possess this property.

If we take the σ_v in question to lie along the $\phi_1 - \phi_4$ line, the antisymmetric orbital can only be

$$\Psi_7 = \frac{1}{\sqrt{2}}(\phi_2 - \phi_3)$$

(5.50)

We now try to use the orbitals (5.48) together with others similar to (5.50) to produce a linear combination which is symmetrical with respect to this σ_v. A process of trial and error yields

$$\Psi_8 = \sqrt{\frac{1}{6}}(2\phi_1 - \phi_2 - \phi_3) \qquad (5.51)$$

which is made up from $(2\phi_1 - \phi_2) - (2\phi_3 - \phi_1) - (\phi_2 - \phi_3)$. The resulting determinant is

$$\begin{vmatrix} x & 0 \\ 0 & x \end{vmatrix} = 0 \qquad (5.52)$$

so that the energy of the orbitals is α, i.e., they are nonbonding, as would be anticipated from the fact that they do not include ϕ_4. It is easily verified that equation (5.49) is satisfied by $x^2 = 0$. The determinant (5.52) is in the diagonal form, so that (5.50) and (5.51) are symmetry orbitals for the doubly degenerate energy level.

5d DEFECTS OF THE SIMPLE THEORY

We shall merely list these defects and indicate how improvements might be made, without considering any specific examples.

(i) The basic assumption which we have made is that a linear combination of atomic orbitals is able to provide a good approximation to the true one-electron wave functions of a molecule. In the simple theory this is mitigated by the fact that α and β are regarded as empirical parameters, rather than being derived from the theory, so that the atomic orbitals are not restricted to being carbon $2p$ orbitals but may be any kind of orbital for which the values assumed for H_{rs} and S_{rs} are reasonable approximations to the correct values. Because of the way in which errors can be absorbed into the empirical parameters, the simple theory is often much more successful than it has any apparent right to be.

(ii) We have specifically neglected interactions between the σ-bond framework of the molecule and the π-electron system. Such interaction can arise in three different ways: (*a*) If the carbon-carbon distance differs for different bonds in the molecule (as in butadiene), the value of β will also differ. (*b*) There may be significant Coulomb repulsion between σ and π charge clouds, especially if either charge distribution is unsymmetrical. Variations in both

α and β can result from the screening effect of the σ charge cloud between the nucleus and the π electrons. (*c*) There may be overlap and hence exchange of electrons between σ and π bond systems, as in hyperconjugation. This is not usually an important effect.

(iii)   The assumption of zero overlap ($S_{rs} = \delta_s^r$), even for orbitals whose overlap is essential if bonding is to occur, is obviously un-realistic. Nevertheless, as indicated initially, the theory is not noticeably improved by leaving out this assumption. Errors arising from this and from the assumption that H_{rs} is zero if atoms r and s are not bonded are also readily taken up by the empirical param-eters α and β.

(iv)   A major source of difficulty in all many-electron problems is electron correlation, i.e., the fact that electrons do not move independently but repel each other and so tend to remain apart as much as possible. (This is the theoretical basis of Hund's rule which we use when carrying out the *aufbau* process: states of high multiplic-ity are of lower energy than states of low multiplicity, because spin-paired electrons are obliged to be relatively close together in the same orbital.) In our calculations we have determined one-electron wave functions and then usually assigned two electrons to them.

There are two common methods of taking electron repulsions into account. The first is to estimate electron repulsion energies for all available orbitals and to include in the ground state wave function appropriately weighted contributions from higher energy orbitals which have less electron repulsion. This is known as allowing for "configuration interaction." The second method is to consider the one-electron Hamiltonian to involve the field of all the other electrons, as well as that of the carbon nuclei, i.e., to use a self-consistent field Hamiltonian. Such calculations are obviously going to be lengthy.

(v)   The aufbau procedure relies on the assumption that the wave function for the molecule as a whole is of the form

$$\Psi = \Psi_1(1) \cdot \Psi_2(2) \cdot \Psi_3(3) \cdots \Psi_n(n) \qquad (5.53)$$

where the numbers in parentheses serve to label the different elec-trons, and the spin coordinate is here included in the one-electron wave functions Ψ_m. Since electrons cannot be labeled in this way, we should also take account of other possible wave functions, such as

$$\Psi = \Psi_1(2) \cdot \Psi_2(1) \cdot \Psi_3(3) \cdots \Psi_n(n) \qquad (5.54)$$

and the general solution of the Schroedinger equation for the molecule must be a linear combination of all possible functions of this sort. This linear combination should also be antisymmetric with respect to the interchange of two electrons, as required by the Pauli exclusion principle. These requirements can be met by using *antisymmetrized molecular orbitals* of the form

$$\Psi = \frac{1}{\sqrt{n!}} \times \begin{vmatrix} \Psi_1(1) & \Psi_2(1) & \Psi_3(1) & \cdots & \Psi_n(1) \\ \Psi_1(2) & \Psi_2(2) & \Psi_3(2) & \cdots & \Psi_n(2) \\ \cdot & \cdot & \cdot & \cdot & \cdot & \cdot \\ \Psi_1(n) & \Psi_2(n) & \Psi_3(n) & \cdots & \Psi_n(n) \end{vmatrix} \qquad (5.55)$$

Calculations of this order of complexity are not commonly carried out by the ordinary organic chemist.

5e MISCELLANEOUS TOPICS

Calculation of Electron Densities. It is obvious from sketches of L.C.A.O. molecular orbitals, such as those in Fig. 11, that to a good approximation the value of the wave function Ψ at atom i is $a_i\phi_i$. Hence the electron density at this atom due to electrons in the orbital Ψ is proportional to $(a_i\phi_i)^2$. If the molecular orbitals are normalized and the number of electrons per orbital is n ($= 0, 1$, or 2), then the total electron density at atom i is given by

$$q_i = \sum n \cdot a_i^2 \qquad (5.56)$$

where the summation is taken over all the occupied molecular orbitals.

For a class of molecules which are known as *alternant hydrocarbons*, the π-electron density at each atom is unity (assuming that each atom contributes one half-filled p-orbital to the π-electron system). Alternant hydrocarbons are planar, conjugated hydrocarbons, having no odd-membered rings, in which the carbon atoms can be divided into two sets, "starred" and "unstarred," such that

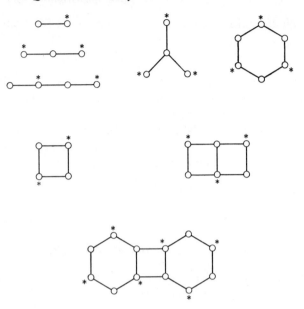

Fig. 12   Some alternant hydrocarbons.

each starred atom is bonded only to unstarred and vice versa. Some examples are shown in Fig. 12.

Bond Order and Free Valence.   The bond order for the π-bonding between atoms i and j is defined as

$$P_{ij} = \sum n \cdot a_i \cdot a_j \qquad (5.57)$$

where n, a_i, and a_j have the same significance as before, and the summation is again taken over all occupied molecular orbitals. The partial bond order (or partial mobile bond order) is simply

$$p_{ij} = a_i \cdot a_j \qquad (5.58)$$

and is positive for a bonding orbital, zero for a nonbonding orbital, and negative for an antibonding orbital between atoms i and j.

The free valence index F_i is defined by

$$F_i = N_{\max} - N_i \qquad (5.59)$$

where N_i is the sum of the orders of the bonds to atom i and $N_{\max}$ is the greatest possible value of N_i for a carbon atoms.   Including σ bonds, the highest N_i value that has been found for carbon is 4.732, in trimethylene methane. The free valence index can be used as an indication of the ease of attack by free radicals at the atom in question.

Experimental Value of β. We have previously derived a value of 2β for the delocalization energy of benzene by comparison of the π-electron energy of the molecule with that of a system of three ethylenic double bonds. This delocalization energy may tentatively be equated to the resonance energy of benzene, which is a quantity derived from valence bond theory and which is equal to the energy difference between the actual benzene molecule and one of the Kekulé structures:

$$\text{Resonance energy} = \text{energy} \left(\vcenter{\hbox{⬡}} \longrightarrow \vcenter{\hbox{⬡}} \right) \qquad (5.60)$$

Resonance energies can be determined empirically in a variety of ways, for example, a value of -63 kcal./mole for the resonance energy of benzene may be deduced from bond energies, based on enthalpies of dissociation, as shown in the following calculation.[5] For methane:

$$CH_4(g) \rightarrow C(g) + 4H(g), \Delta H_{298} = 397.2 \text{ kcal.};$$
$$\therefore \quad E(\text{C—H}) = 99.3 \text{ kcal.}$$

while for ethane:

$$C_2H_6(g) \rightarrow 2C(g) + 6H(g), \Delta H_{298} = 674.6 \text{ kcal.};$$
$$\therefore \quad E(\text{C—C}) = 79.2 \text{ kcal.}$$

and for ethylene:

$$C_2H_4(g) \rightarrow 2C(g) + 4H(g), \Delta H_{298} = 537.7 \text{ kcal.};$$
$$\therefore \quad E(\text{C}{=}\text{C}) = 140.6 \text{ kcal.}$$

For benzene it is found by experiment that

$$C_6H_6(g) \rightarrow 6C(g) + 6H(g), \Delta H_{298} = 1318.1 \text{ kcal.}$$

For one of the Kekulé structures we would expect the enthalpy of dissociation to be $6E(\text{C—H}) + 3E(\text{C—C}) + 3E(\text{C}{=}\text{C}) = 1255.2$ kcal., and the difference between this and the observed value is -62.9 kcal./mole, corresponding to $\beta = -31$ kcal./mole. We note

---

[5] These calculations are based on heats of formation which are given in Appendix 7 of Lewis, Randall, Pitzer, and Brewer, *Thermodynamics*, New York: McGraw-Hill, 1961.

that $E(C—H)$ is assumed to be the same in ethylene as in methane, although in one case the bonding is with sp^3 hybrid orbitals of carbon and in the other sp^2 hybrids are used.

Similar calculations based on heats of combustion and on heats of hydrogenation both yield the same value, namely -36 kcal./mole, for the resonance energy, corresponding to $\beta = -18$ kcal./mole. Several other correlations exist which yield values of β for aromatic systems in the -16 to -20 kcal./mole range.   In every case the resonance energy is obtained as the small difference between two large quantities and this degree of consistency is probably to be regarded as satisfactory.

The *vertical resonance energy* is defined as the energy difference between the benzene molecule and a Kekulé-type structure having bonds equal in length to those in benzene.   An actual Kekulé structure would, of course, have alternating bond lengths.   The vertical resonance energy is expected to correspond more closely to the calculated delocalization energy than do the empirical resonance energies already considered, hence it is of interest to attempt to estimate the *distortion energy* which would be required to equalize the bond lengths in a Kekulé structure.   This distortion energy has been variously calculated, with resulting vertical resonance energies of -63, -73 and -111.5 kcal./mole, so it would appear that the best value of β is about -35 kcal./mole.   A "spectroscopic value" of β of -55 kcal./mole has been derived by Platt[6] from comparisons of calculated and observed energy level separations in aromatic hydrocarbons, and this higher value appears to be the most suitable one for applications of the theory to spectroscopy.

EXERCISES

5.1 Write out the secular determinant in terms of $x = (\alpha - E)/\beta$ for each of the following molecular skeletons:

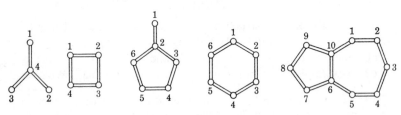

[6] J. R. Platt, *J. Chem. Phys.*, **15**, 419 (1947).

5.2 Complete the calculation for butadiene in section 5c, p. 84.

5.3 Calculate the delocalization energy of bicyclohexatriene (C_{2v}).

5.4 Sketch the molecular orbitals of benzene from equations (5.36) and (5.39). Indicate the presence of nodes in the wave functions, and correlate the number of nodes (regions where the wave function is zero, in this case they are nodal planes) with the degree of bonding or anti-bonding character of the orbitals.

5.5 Work out the energy levels for ethylene using $S_{12} = 0.25$. Repeat for butadiene with this value of the overlap integral. (We now have $x = (\alpha - E)/(\beta - 0.25E)$.) Compare the delocalization energy with that which was obtained in section 5c, p. 80.

5.6 Calculate the π-bond orders and free valence indices for butadiene in the ground state and with one electron excited from the highest bonding orbital to the lowest antibonding orbital.[7]

5.7 Calculate delocalization energies, π-bond orders, and free valence indices for cyclopentadienyl anion(I), for fulvene(II), and for cycloheptatrienyl cation(III).

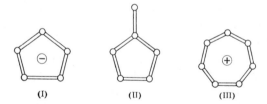

(I) (II) (III)

5.8 Calculate the dipole moment of pyrrole, assuming $\int \phi_c \hat{H} \phi_N \, d\tau = k\beta$ for carbon bonded to nitrogen, and $\int \phi_N \hat{H} \phi_N \, d\tau = \alpha + h\beta$, where α and β are the corresponding integrals for the carbon atom and k and h take the values 1.0 and 0.5, respectively.[8] The experimental value of the dipole moment is $+1.78D$.

---

[7] Complete calculations for the ground state configuration are given by Roberts in *Notes on Molecular Orbital Calculations*, New York: W. A. Benjamin, 1961.

[8] See Coulson and Longuet-Higgins, *Trans. Faraday, Soc.*, **43**, 87 (1947) and Coulson, *Valence*, 2nd Edition, London: Oxford University Press, 1961, p. 258.

6

Ligand Field Theory

....

6a INTRODUCTION

Ligand field theory has been developed as a means of calculating wave functions and energy levels for atoms in complexes. Here we understand by a complex any entity which consists of a central atom or ion surrounded by a fixed arrangement of two or more ligands. This definition includes not only the usual coordination complexes but also the case of an atom surrounded by its nearest neighbors in a crystal lattice, and many stable species such as SO_4^{2-}.

The most interesting applications of the theory are to complexes of atoms or ions which possess incomplete d or f electron shells, i.e., to complexes of metals of the transition and inner transition series of the periodic table. In a free atom a set of five d or seven f orbitals is perfectly degenerate, but in a complex these sets of orbitals are obliged to split into smaller sets, each of which transforms according to some irreducible representation of the point group of the complex. Thus, for example, the largest irreducible representations of the octahedral point group O_h are of dimension 3, so that no energy level in a complex of this symmetry can be more than three-fold degenerate. The various sets of orbitals differ in their energy of interaction with the field of the ligands, and the aim of the theory is first to determine the manner in which the orbitals are distributed among the different irreducible representations and then to calculate the corresponding energy levels.

Ligand field theory contains as limiting cases the crystal field theory of "ionic" complexes and the molecular orbital theory of strongly bound, "covalent" complexes. The basic approach, which is common to the three theories, is to reduce the representation of the point group of the complex for which the set of d or f orbitals forms a basis, and to determine the nature of the atomic orbitals which comprise the resulting irreducible representations. In crystal field theory the energies of these orbitals are then calculated on the assumption that the interaction of the metal electrons with the field of the ligands is entirely electrostatic, and in favorable cases order-of-magnitude agreement with experiment may be obtained. In ligand field theory similar calculations are carried out, but here it is recognized that the orbitals in question are not likely to be pure d or f orbitals of the metal atom but will have varying amounts of ligand orbital character because of overlap of the electron systems. In consequence it is no longer possible to calculate the energies directly, and the results are expressed in terms of empirical parameters, such as the effective charge or dipole of the ligand and the metal-ligand distance. In the molecular orbital theory the metal orbitals of appropriate symmetry are combined with orbitals of the ligands to form molecular orbitals for the complex as a whole. The resulting molecular orbitals may be of σ or π type, depending on whether the metal orbitals are directed toward or between the ligands, and it is usual to assume initially that π bonding is negligible. In the present state of the theory it is not possible to calculate the energies of these orbitals with any degree of precision, and the main use of the molecular orbital theory is to provide a chemically realistic picture of the manner in which the energy levels in a complex are determined by the bonding of the ligands. In this account of ligand field theory we shall not attempt any actual calculations of energies, but will content ourselves with a consideration of that which is common to all three theories, the determination of the splitting of orbital degeneracy in the presence of a symmetrical ligand field. Before we can understand the properties of atoms in complexes, however, we must first be familiar with the quantum theory of a free atom or ion, and an introductory account of this theory is given in the next section.

6b THE FREE ATOM

The Hamiltonian operator for a many-electron atom has the form

$$\hat{H} = \frac{-\hbar^2}{2m} \sum_j \nabla_j^2 - \sum_j \frac{Z_j e^2}{r_j} + \sum_{j>k} \frac{e^2}{r_{jk}} + \zeta_j \sum_j \mathbf{l}_j \cdot \mathbf{s}_j \qquad (6.1)$$

In this expression ∇_j^2 is the Laplacian operator for the coordinates of electron j, $Z_j e$ is the effective nuclear charge for electron j (this makes it possible to allow for screening of the nucleus by other electrons), r_j is the distance of electron j from the nucleus, r_{jk} is the distance between electrons j and k (the term in r_{jk} takes account of the electrostatic repulsion between different electrons; the summation is written in such a way that each repulsion term is counted only once), and the final term takes account of the magnetic interaction between the spin and orbital angular momenta of the electrons, i.e., of spin-orbit coupling. Here $\mathbf{l}_j$ and $\mathbf{s}_j$ are the orbital and spin angular momentum vectors while ζ_j is a quantity which depends on the potential gradient to which electron j is subjected, and so is a function of r_j.

The presence of the term in r_{jk} makes it impossible to separate the variables in the Schroedinger equation, so that approximate methods of calculation must be used. The usual approach is to take

$$\hat{H}^0 = \frac{-\hbar^2}{2m} \sum_j \nabla_j^2 - \sum_j \frac{Z_j e^2}{r_j} \qquad (6.2)$$

as the Hamiltonian operator of an "unperturbed system" and to apply

$$\hat{H}' = \sum_{j>k} \frac{e^2}{r_{jk}} \qquad (6.3)$$

as a perturbation. The spin-orbit coupling term is in general much smaller than the others and can be added in later.

The unperturbed Hamiltonian (6.2) immediately separates into a series of one-electron Hamiltonians

$$\hat{H}_j^0 = \frac{-\hbar^2}{2m} \nabla_j^2 - \frac{Z_j e^2}{r_j} \qquad (6.4)$$

and the corresponding Schroedinger equations are identical with that for a hydrogen-like atom, as discussed in section 2e. Hence our unperturbed one-electron wave functions are of the form

$$\Psi(n, l, m) = R_{nl}(r) \cdot Y_{lm}(\theta, \phi) \qquad (6.5)$$

where the detailed expressions for $R_{nl}(r)$ and $Y_{lm}(\theta, \phi)$ are given in section 2e. We shall not require these expressions here, but merely recall that the principal quantum number is an integer and may range from one to infinity, the azimuthal quantum number l takes all integral values from zero to $n - 1$, and the magnetic quantum number m takes all integral values from l to $-l$. The physical significance of l and m is that $l(l + 1)\hbar^2$ is the square of the orbital angular momentum of the electron, and $m\hbar$ is the component of the orbital angular momentum in a particular direction, usually designated as the z direction. The orbitals are labeled in the usual way, with the principal quantum number and with a small letter s, p, d, f, g, h, i, $\cdots$ for l equal to 0, 1, 2, 3, 4, 5, 6, $\cdots$.

A state of an atom in which the number of electrons in each orbital is specified is called a "configuration." Thus, for example, the ground state configuration of the nitrogen atom is $1s^2 2s^2 2p^3$.

In practice it is common to use for the radial factor in (6.5) the approximate expression (*Slater orbital*):

$$R(r) = N \cdot r^{(n^* - 1)} \cdot e^{-(Z - s)r/n^* a_0} \qquad (6.6)$$

where n^* and s are adjustable constants and N is a normalizing factor. Here n^* is known as the effective quantum number and s is a screening parameter. The values of these parameters are fixed in accordance with *Slater's rules*, which are as follows:

1. n^* depends on the value of the principal quantum number, as shown in the following table:

Value of n:	1	2	3	4	5	6
Value of n^*:	1	2	3	3.7	4.0	4.2

2. The electron shells are divided into the following groups for the purpose of deciding the value of s:

$$1s; \ 2s, \ 2p; \ 3s, \ 3p; \ 3d; \ 4s, \ 4p; \ 4d, \ 4f; \ \cdots$$

3. The screening constant s is made up from the following contributions:

    (*a*) Nothing from any shell outside the one being considered.

    (*b*) An amount 0.35 is added for every other electron in the same group, except that for the $1s$ group an amount 0.30 is added.

(c) If the shell considered is s or p, an amount 0.85 is added for each electron with principal quantum number less by one, and 1.00 for each electron further in; if the shell is d or f, the amount is 1.00 for each inner electron.

In building up the complete wave function of an atom as a product of one-electron functions it is necessary to take account of electron spin and the Pauli exclusion principle. A typical product wave function with spin included is

$$\Psi(1, 0, 0, \tfrac{1}{2}/1) \cdot \Psi(1, 0, 0, \bar{\tfrac{1}{2}}/2) \cdots \Psi(n, l, m, s/k) \qquad (6.7)$$

where n, l, m, and s are the quantum numbers of electron k and $\bar{\tfrac{1}{2}}$ stands for $-\tfrac{1}{2}$. This sort of product wave function does not automatically comply with the Pauli exclusion principle, which requires that it be antisymmetric with respect to the interchange of any two electrons. A function which does fulfill this requirement is the Slater determinant

$$\Psi = \frac{1}{\sqrt{k!}} \times \begin{vmatrix} \Psi(1, 0, 0, \tfrac{1}{2}/1) & \Psi(1, 0, 0, \tfrac{1}{2}/2) & \cdots & \Psi(1, 0, 0, \tfrac{1}{2}/k) \\ \Psi(1, 0, 0, \bar{\tfrac{1}{2}}/1) & \Psi(1, 0, 0, \bar{\tfrac{1}{2}}/2) & \cdots & \Psi(1, 0, 0, \bar{\tfrac{1}{2}}/k) \\ \cdot \cdot \cdot \cdot & \cdot \cdot \cdot \cdot & \cdot \cdot \cdot & \cdot \cdot \cdot \\ \Psi(n, l, m, s/1) & \Psi(n, l, m, s/2) & \cdots & \Psi(n, l, m, s/k) \end{vmatrix}$$

$$(6.8)$$

The function (6.8) has been normalized on the assumption that the $\Psi(n, l, m, s)$ form an orthonormal set. The determinant may be expanded to yield a sum of all possible wave functions similar to (6.7). By the properties of determinants, this sum changes sign if two columns, corresponding to two different electrons, are interchanged, and reduces to zero if two rows are identical, i.e., if two electrons have the same set of quantum numbers n, l, m, and s. We shall not perform any actual calculations of configuration energies, but if we were to do so it would be necessary to evaluate matrix elements of the Hamiltonian operator in terms of antisymmetrized atomic orbitals similar to (6.8). More accurate wave functions which include allowance for electron repulsion terms may be calculated by the "Hartree-Fock self-consistent field method." Details of such calculations may be found in several of the books which are listed in Appendix I.

We noticed in Chapter 2 that the angular momentum operators $\hat{M}^2$ and $\hat{M}_z$ commute with the Hamiltonian, so that we can expect the states of an atom to be characterized by particular values of angular momentum as well as by their energies. A given electron configuration may give rise to several different values of the total angular momentum, and in the unperturbed system which we have been considering these different angular momenta all correspond to the same total energy. Under the influence of the perturbation $\hat{H}'$, however, this degeneracy is lifted, and the value of the total angular momentum then provides a useful label with which to distinguish a state from others which arise from the same configuration but which differ in their electron repulsion energies. The situation is somewhat complicated by the fact that each electron in an atom possesses both a spin and an orbital angular momentum, and it is usually necessary to specify the order in which the addition of vectors is carried out. There are in fact two important ways in which the angular momentum vectors may be combined, one of which is known as "Russell-Saunders" (or $L - S$) coupling and the other as "$j - j$ coupling."

In Russell-Saunders coupling, which is the most important case, the orbital angular momentum vectors **l** of the different electrons combine to produce a resultant **L**, and the spin vectors **s** similarly combine to produce a resultant **S**. The vectors **L** and **S** then combine to produce a resultant **J** for the whole atom. The resultants are found by the rule which was obtained in section 1e, namely that two angular momentum vectors $\mathbf{l}_1$ and $\mathbf{l}_2$, of magnitude $\sqrt{l_1(l_1 + 1)} \cdot \hbar$ and $\sqrt{l_2(l_2 + 1)} \cdot \hbar$, form a resultant **L** of magnitude $\sqrt{L(L + 1)} \cdot \hbar$, where L takes any of the values $l_1 + l_2, l_1 + l_2 - 1, \cdots l_1 - l_2$ ($l_1 \geqslant l_2$). For each value of L there is a further quantum number M_L which takes any of the $2L + 1$ values $L, L - 1, L - 2, \cdots - L$, and a similar spin quantum number M_S takes the values $S, S - 1, \cdots - S$. Here $M_L \hbar$ and $M_S \hbar$ are the components of angular momentum in a particular direction, i.e., in the terms of section 1e they give the magnitude of $\mathbf{L}_z$ and $\mathbf{S}_z$, respectively. The quantum number J for the total angular momentum of the atom takes one of the values $L + S, L + S - 1, \cdots L - S$ for $L \geqslant S$, or $S + L, S + L - 1, \cdots S - L$ for $S > L$.

In Russell-Saunders coupling the different states of the atom are labeled S, P, D, F, G, H, I, $\cdots$ according to whether L has the value 0, 1, 2, 3, 4, 5, 6, $\cdots$ respectively. The "multiplicity" of a state, i.e., the number of different J values, is equal to $2S + 1$ for $L \geqslant S$, and is written as a superscript in front of the capital letter which designates the state, for example, ^2P or ^2D. If S is greater than L, the actual multiplicity is only $2L + 1$, but by convention the value of $2S + 1$ is still used for the superscript, for example, ^4S. A symbol such as ^4S ("quartet S") or ^2P ("doublet P") is called a "spectroscopic term." The value of J for each component state of a term is written as a subscript following the capital letter. Thus for ^2D, ^2P, and ^4S terms we obtain the states $^2D_{5/2}$ (doublet D five-halves), $^2D_{3/2}$ (doublet D three-halves), $^2P_{3/2}$, $^2P_{1/2}$, and $^4S_{3/2}$. In addition to these symbols it is common to label odd states, i.e., those for which the total wave function is antisymmetric with respect to inversion in the nucleus, with a small superscript 0 after the capital letter. Alternatively the letters g (gerade) and u (ungerade) may be used as subscripts for even and odd states, respectively. A state is odd or even depending on whether $\sum_k l_k$ for all electrons is odd or even.

The foregoing method of labeling atomic states is very often used even when $j - j$ coupling occurs. In $j - j$ coupling the individual **l** and **s** of each electron combine to form a resultant **j**, and the various **j** then combine to form a resultant **J** for the whole atom. A state is labeled simply with the value of the quantum number J. This type of coupling is approached with very heavy atoms and with highly excited states of light atoms. In such cases the "multiplet splitting" between states with different values of J is comparable with the energy differences between pairs of terms when the labeling is carried out in accordance with Russell-Saunders coupling. Normally, in Russell-Saunders coupling, the multiplet splitting is small in comparison with the energy differences between terms.

As mentioned before, in the absence of the electron repulsions described by $\hat{H}'$ the different terms arising from a single configuration would all have the same energy. The introduction of $\hat{H}'$ removes this degeneracy, after which the identity of the lowest term may be deduced with the aid of Hund's rule. According to Hund's rule the lowest term of a given configuration is that which has the greatest multiplicity, while of two terms with the same multiplicity the lowest is that with the greatest value of L. This rule also provides a useful though not infallible guide to the order of terms which lie above the lowest state.

To the degree of approximation which includes only $\hat{H}^0$ and $\hat{H}'$ in the Hamiltonian, the component states of a term all have the same energy, thus $^2D_{5/2}$ and $^2D_{3/2}$ coincide. This degeneracy disappears when the spin-orbit coupling term of equation (6.1) is introduced as a further perturbation. States for which J is greater than zero retain a $(2J + 1)$-fold degeneracy, corresponding to the number of possible values of the quantum number M_J, and in a magnetic field the states are observed to split into this number of components (*Zeeman effect.*) Finally, in addition to this structure each energy level may possess a "hyperfine" structure owing to interaction of **J** with the nuclear spin of the atom. The hyperfine structure is not apparent in ordinary experiments and can usually be neglected. The energy levels which arise from the ground state configuration as the different perturbation terms are added are shown in Fig. 13 for the case of atomic nitrogen.

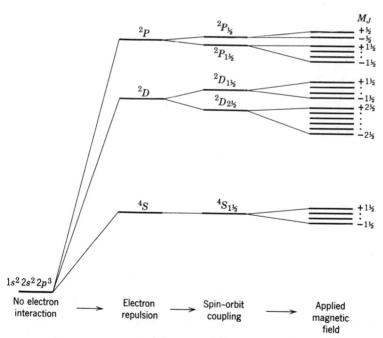

Fig. 13   Energy levels of atomic nitrogen (not to scale).

It remains to consider how we can decide which terms arise from a given configuration. The calculation is simplified by the fact, which is easily verified, that filled shells such as s^2, p^6, d^{10}, or f^{14} always possess zero orbital and spin momenta because of the Pauli principle. This means that we have only to consider contributions from electrons which occupy incomplete shells. If there are no equivalent electrons, i.e., no two electrons occupy the same set of s, p, d, or f orbitals, the resultant angular momenta are found simply by vector addition. Thus a configuration (filled shells) $p^1 d^1$ gives rise to singlet and triplet P, D, and F terms. If equivalent electrons are involved, it is necessary to take account of the Pauli exclusion principle, and to illustrate the procedure in this case we consider a nitrogen atom, for which the configuration is p^3.

An electron in a p-type atomic orbital, where $l = 1$, can have m_l equal to 1, 0, or -1, and m_s equal to $\frac{1}{2}$ or $-\frac{1}{2}$, but no two electrons may have the same values of both m_l and m_s. Hence, we can build up the final state of the atom by combining any three of the following (m_l, m_s) states for individual electrons:

$$(1, \tfrac{1}{2}), (1, -\tfrac{1}{2}), (0, \tfrac{1}{2}), (0, -\tfrac{1}{2}), (-1, \tfrac{1}{2}), (-1, -\tfrac{1}{2}) \qquad (6.9)$$

In effect we are now forming simple product wave functions similar to (6.7). The results may be set out as in Table 6.1. It is not necessary to write out the results for negative values of M_L and M_S since it is obvious that the table is symmetrical about $M_L = M_S = 0$.

Table 6.1

	$M_S = 3/2$	$M_S = 1/2$
$M_L = 2$		$(1, \tfrac{1}{2}) \cdot (1, -\tfrac{1}{2}) \cdot (0, \tfrac{1}{2})$
$M_L = 1$		$(1, \tfrac{1}{2}) \cdot (1, -\tfrac{1}{2}) \cdot (-1, \tfrac{1}{2})$ $(1, \tfrac{1}{2}) \cdot (0, \tfrac{1}{2}) \cdot (0, -\tfrac{1}{2})$
$M_L = 0$	$(1, \tfrac{1}{2}) \cdot (0, \tfrac{1}{2}) \cdot (-1, \tfrac{1}{2})$	$(1, \tfrac{1}{2}) \cdot (0, \tfrac{1}{2}) \cdot (-1, -\tfrac{1}{2})$ $(1, \tfrac{1}{2}) \cdot (0, -\tfrac{1}{2}) \cdot (-1, -\tfrac{1}{2})$ $(1, -\tfrac{1}{2}) \cdot (0, \tfrac{1}{2}) \cdot (-1. \tfrac{1}{2})$

The spectroscopic terms may be extracted from Table 6.1 as follows. The highest value of M_L is 2, with $M_S = 1/2$, so there must be a ^2D term. This term will give rise to one entry (just which entry does not matter) in each of the spaces $M_S = 1/2$ and $M_L = 2$, 1, 0, -1, -2 (the last two not shown in Table 6.1). When these have been eliminated the next highest value of M_L corresponds to a ^2P term, and so we must remove an entry in each of the spaces $M_S = 1/2$; $M_L = 1$, 0, -1. The remaining entries in the table are then recognized as the Zeeman components of a ^4S term, and by Hund's rule the order of the terms is as shown in Fig. 13. This would be a good point at which to attempt Exercise 6.1.

We can similarly work out the terms which arise from any other configuration involving equivalent electrons. The results for configurations of equivalent p and d electrons are listed in Table 6.2. Where a configuration gives rise to several terms of the same type the number of each type is written in brackets after the term symbol.

Table 6.2   Terms for Configurations of Equivalent Electrons

p^0, p^6	^1S
p, p^5	^2P
p^2, p^4	^1S, ^1D, ^3P
p^3	^2P, ^2D, ^4S
d^0, d^{10}	^1S
d, d^9	^2D
d^2, d^8	^1S, ^1D, ^1G, ^3P, ^3F
d^3, d^7	$^2P(2)$, ^2G, ^2F, ^2H, ^4P, ^4F
d^4, d^6	$^1S(2)$, $^1D(2)$, ^1F, $^1G(2)$, ^1I, $^3P(2)$, ^3D, $^3F(2)$, ^3G, ^3H, ^5D
d^5	^2S, ^2P, $^2D(3)$, $^2F(2)$, $^2G(2)$, ^2H, ^2I, ^4P, ^4D, ^4F, ^4G, ^6S

The most obvious feature of Table 6.2 is that the same terms arise from a configuration "filled shells plus n equivalent electrons" as from the configuration "filled shells minus n equivalent electrons." This is because a shell which lacks a certain number of electrons is formally equivalent to one containing the same number of positrons. The order of the terms is still as given by Hund's rule because positron-positron repulsions have the same effect as electron-electron repulsions. The effects of spin-orbit coupling, on the other hand, are in the opposite direction for positrons, and in general for shells which are less than half-filled, the energy of the components of a term increases with increasing J, while for shells which are more than half-filled the energy decreases with increasing J. For an atom with a half-filled shell the multiplet splitting is therefore expected to be very small, and in fact for nitrogen the energies are $^4S_{3/2}$:0; $^2D_{5/2}$:19,223 cm^{-1}; $^2D_{3/2}$:19,231 cm^{-1}; $^2P_{1/2}$, and $^2P_{3/2}$ both 28,840 cm^{-1}.[1]

Selection rules for radiative transitions between atomic energy levels are discussed in Chapter 7.

6c   THE ATOM IN A COMPLEX

The Hamiltonian operator for an atom in a complex has the form

$$\hat{H} = \frac{-\hbar^2}{2m} \sum_j \nabla_j^2 - \sum_j \frac{z_j e^2}{r_j} + \sum_{j>k} \frac{e^2}{r_{jk}} + \zeta_j \sum_j \mathbf{l}_j \cdot \mathbf{s}_j + V \quad (6.10)$$

which is identical, except for the last term, with the Hamiltonian for a free atom.   In this expression the last term V represents the energy of interaction with the ligands.   The type of situation to which this leads depends on the relative magnitudes of V and the other two terms which we have so far considered as perturbations. We can distinguish three different limiting cases as follows:

(i) $V < \zeta_j \sum_j \mathbf{l}_j \cdot \mathbf{s}_j$ Complexes of the rare earths.

(ii) $\sum_{j>k} \frac{e^2}{r_{jk}} > V > \zeta_j \sum_j \mathbf{l}_j \cdot \mathbf{s}_j$ Complexes of the first transition series.

(iii) $V > \sum_{j>k} \frac{e^2}{r_{jk}}$ "Covalent" complexes.

The second and third cases are often called the weak and strong crystalline field cases, respectively.   By comparison with the situation which is illustrated in Fig. 13 we deduce that in case (i) the effect of the field is to remove the degeneracy associated with the different values of M_J for any state; in case (ii) the nature of the

---

[1] Energy values for excited states of a large number of neutral and ionized atoms are listed in "Atomic Energy Levels," by C. E. Moore, *Circular 467 of the National Bureau of Standards*, Washington D.C., Volume I, *Hydrogen Vanadium*; Volume II, *Chromium–Niobium*; Volume III, *Molybdenum–Lanthanum* and *Hafnium–Actinium*.

states which arise from a particular term is altered; and in case (iii) the terms themselves lose their significance.

The potential V may be separated into a spherically symmetrical part V_R and a part V' which has the symmetry of the arrangement of ligands. Of these terms, V_R is generally much the greater and is responsible for most of the heat of formation of a complex or the lattice energy of a crystal. The effect of this term on the energy levels is to shift the whole system bodily downwards, without removing any degeneracies or altering the overall pattern. The intervals between terms are generally reduced by about one-third in a complex owing to the reduced importance of interelectron repulsions in the presence of the pervading field V_R.

The term V', although smaller than V_R, has more interesting experimental consequences because of its symmetry properties. In a spherically symmetrical environment p, d, and f atomic orbitals possess respectively 3-, 5-, and 7-fold degeneracies. Under the influence of V' the formerly degenerate orbitals become segregated into sets, each of which transforms according to an irreducible representation of the point group of V'. Within a set the energy of the orbitals remains constant but the effect of the potential V' varies from one set to another, and the original degeneracy is partially or wholly removed. Thus the whole pattern of the energy levels is altered and this has many important consequences, particularly with regard to the spectroscopic and magnetic properties of the complex.

In the remainder of this section we shall see how it is possible to use group theory to determine the manner in which a set of degenerate orbitals is split up by interaction with the potential V'. Our main concern will be to find the effect of a nonspherical potential V' on terms arising from configurations of equivalent d electrons. Group theory alone cannot tell us anything about the energies of the resulting states, and the consideration of energy level diagrams is deferred until the next section.

Let us consider the effect of an octahedral field on a set of five degenerate d orbitals. The complete octahedral point group O_h is obtained by adding the operation of inversion, i, to the pure rotations E, C_2, C_3, and C_4 which comprise the point group O. This has the effect of doubling the number of irreducible representations, since each representation of O now has the opportunity to be symmetric or antisymmetric (*g* or *u*) with respect to the operation of inversion in the center of symmetry. It is easier for our purposes to deal with O rather than with O_h, and we shall do this, remembering that s, d, g, $\cdots$ orbitals can give rise only to g representations while p, f, h, $\cdots$ orbitals can give rise only to u representations, because of the g or u nature of the individual orbitals. Our aim, therefore, is to find the characters of the transformation matrices which describe the effects of pure rotations on a set of d orbitals.

The *d*-orbital functions are of the form

$$\Psi = R(r) \cdot Y_{2m}(\theta, \phi) \tag{6.11}$$

where m takes the values 2, 1, 0, -1, -2, and where $Y_{2m}(\theta, \phi)$ is of the form

$$Th(\theta) \cdot \frac{1}{\sqrt{2\pi}} \cdot e^{im\phi} \tag{6.12}$$

We recall that $m\hbar$ is the component of orbital angular momentum in the z direction. There is nothing to prevent us defining the z axis of the atom as the axis about which the rotations are carried out, and this leads to a considerable simplification because θ is invariant with respect to rotations about this axis. The radial function R is already invariant with respect to all the operations of a point group; hence we have only to consider the matrices which describe the effects of rotations on the functions $e^{im\phi}$. A rotation by an angle α about the z axis changes $e^{im\phi}$ to $e^{im(\phi + \alpha)}$, so the matrix equation for the transformation is

$$\begin{pmatrix} e^{2i(\phi+\alpha)} \\ e^{i(\phi+\alpha)} \\ e^{0} \\ e^{-i(\phi+\alpha)} \\ e^{-2i(\phi+\alpha)} \end{pmatrix} = \begin{pmatrix} e^{2i\alpha} & 0 & 0 & 0 & 0 \\ 0 & e^{i\alpha} & 0 & 0 & 0 \\ 0 & 0 & e^{0} & 0 & 0 \\ 0 & 0 & 0 & e^{-i\alpha} & 0 \\ 0 & 0 & 0 & 0 & e^{-2i\alpha} \end{pmatrix} \begin{pmatrix} e^{2i\phi} \\ e^{i\phi} \\ e^{0} \\ e^{-i\phi} \\ e^{-2i\phi} \end{pmatrix} \tag{6.13}$$

The character of the operation is, therefore

$$\chi(\alpha) = e^{2i\alpha} + e^{i\alpha} + e^{0} + e^{-i\alpha} + e^{-2i\alpha} \tag{6.14}$$

$$= 2 \cos 2\alpha + 2 \cos \alpha + 1 \tag{6.15}$$

using $\qquad e^{ix} = \cos x + i \sin x$

Hence $\chi(E) = 5$

$$\chi(C_2) = 1 \tag{6.16}$$

$$\chi(C_3) = -1$$

and $\chi(C_4) = -1$

for α equal to 0, π, $2\pi/3$, and $\pi/2$, respectively. By means of equation (4.43) we find

$$\Gamma_d = e_g + t_{2g} \tag{6.17}$$

where small letters are customarily used for the representations to show that they arise from one-electron orbitals (s, p, d, $\cdots$) rather than from atomic terms (S, P, D, $\cdots$). Equation (6.17) was obtained by using the character table for point group O and the g subscripts were added later. We have now shown that in an octahedral field the d orbitals split into a doubly degenerate set and a triply degenerate set which transform according to the e_g and t_{2g} representations, respectively.

The preceding calculation is easily generalized to include other types of orbitals. The character for a rotation by α is in general

$$\chi(\alpha) = e^{il\alpha} + e^{i(l-1)\alpha} + \cdots + e^{-il\alpha} \tag{6.18}$$

where $l = 0, 1, 2, 3, \cdots$ for s, p, d, f, $\cdots$ orbitals. It can be shown that for $\alpha \neq 0$, equation (6.18) becomes

$$\chi(\alpha) = \frac{\sin(l + \frac{1}{2})\alpha}{\sin \frac{1}{2}\alpha} \tag{6.19}$$

while for $\alpha = 0$ we have

$$\chi(E) = 2l + 1 \tag{6.20}$$

We can also use these results to account for the effects of a symmetrical potential on an atomic term. The overall wave function for an atom in Russell-Saunders coupling similarly contains a factor $e^{iM_L\phi}$, and this factor is the only part of the wave function to be affected by a rotation about the z axis. Hence if we consider a D term, we shall find that in an octahedral field it separates into two parts which transform according to the E_g and T_{2g} representations of O_h. It is found that the spin multiplicity of a term is not affected by the field of the ligands, so that, for example, a ^3D term separates into ^3E_g and $^3T_{2g}$ terms in an octahedral field.

The splitting of atomic terms in fields of symmetry O_h, T_d, and D_{4h} is summarized in Table 6.3. The table which gives the effects of these symmetrical fields on atomic orbitals is the same except that the capitals designating atomic terms are everywhere replaced by the small letters which are used to designate one-electron orbitals.

Table 6.3

Term	O_h	T_d	D_{4h}
S	A_{1g}	A_1	A_{1g}
P	T_{1u}	T_2	$A_{2u}+E_u$
D	E_g+T_{2g}	$E+T_2$	$A_{1g}+B_{1g}+B_{2g}+E_g$
F	$A_{2u}+T_{1u}+T_{2u}$	$A_2+T_1+T_2$	$A_{2u}+B_{1u}+B_{2u}+2E_u$
G	$A_{1g}+E_g+T_{1g}+T_{2g}$	$A_1+E+T_1+T_2$	$2A_{1g}+A_{2g}+B_{1g}+B_{2g}+2E_g$
H	$E_u+2T_{1u}+T_{2u}$	$E+T_1+2T_2$	$A_{1u}+2A_{2u}+B_{1u}+B_{2u}+3E_u$
I	$A_{1g}+A_{2g}+E+T_{1g}+2T_{2g}$	$A_1+A_2+E+T_1+2T_2$	$2A_{1g}+A_{2g}+2B_{1g}+2B_{2g}+3E_g$

6d   ENERGY LEVEL DIAGRAMS

So far we have not paid any attention to the relative energies of the orbitals which transform according to the different representations of the point group of a complex. We need to know at least the order of the energies to decide, for example, whether the lowest configuration of a d^3 atom in a strong octahedral field is t_{2g}^3 or e_g^3. The first of these would give rise to a quartet ground state, the second to a doublet.

Calculation of absolute energies in a case like this is very difficult, but the order of the energy levels may be deduced quite easily. Once this is done we can construct a correlation diagram to show qualitatively how the atomic terms are split and displaced as the strength of the field increases, and from the correlation diagram it is possible to make a number of interesting predictions for comparison with experiment. We deduce the order of the energy levels as follows:

Individual d orbitals are labeled d_{xy}, d_{yz}, d_{xz}, $d_{x^2-y^2}$, and d_{z^2}, indicating that the lobes of maximum electron density lie along the coordinate axes for $d_{x^2-y^2}$ and d_{z^2} and between the axes for the others. We can see qualitatively how the orbitals separate into two sets of

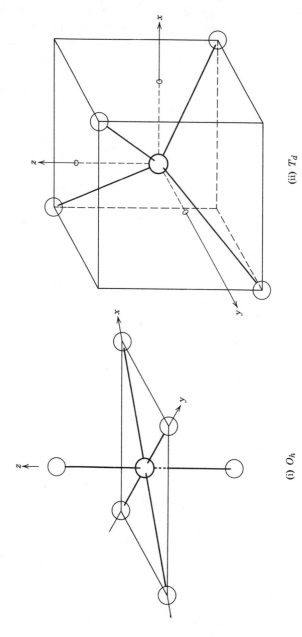

(i) O_h

(ii) T_d

Fig. 14   Relationship of ligand fields of symmetry O_h and T_d to coordinate axes.

different energy in octahedral and tetrahedral complexes by considering the stereochemistry of these complexes in relation to the directions of the lobes of the d orbitals, as shown in Fig. 14.

The ligands are normally either negative ions or dipolar molecules which are oriented so that the negative pole is directed toward the central cation; therefore the energies of d orbitals which point towards the ligands will be increased relative to those which point between the ligands.   Hence we expect $d_{x^2-y^2}$ and d_{z^2} to lie above d_{xy}, d_{yz}, and d_{xz} in an octahedral complex and below them in a tetrahedral complex.   We can tentatively identify the first two as the components of e_g and e and the other three as the components of t_{2g} and t_2 in the point groups O_h and T_d, and a more rigorous derivation shows that this is indeed a correct picture.   The results are illustrated in Fig. 15.   We note that the effect of V_R is omitted from the diagram, and so the center of gravity of the energy level system is not displaced by the field.   The difference between the two sets of energy levels is labeled Δ_0 for the octahedral case and Δ_t for the tetrahedral case.   An alternative label for this quantity is $10D_q$. It can be shown that in the crystal field approximation Δ_t equals $\frac{4}{9}\Delta_0$ for a given ligand.

We will now consider a simple example, namely a d^2 ion (for example, V^{3+}) in an octahedral environment, and construct a

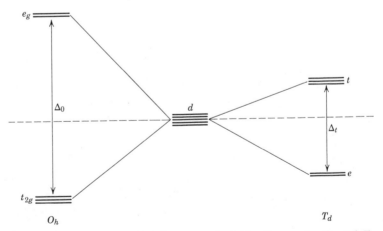

Fig. 15   Splitting of d-orbital degeneracy in fields of symmetry O_h and T_d.

correlation diagram to show how the energy levels of the ion vary with increasing strength of the ligand field.

The free ion terms for the d^2 configuration are given in Table 6.2, and the terms into which they separate under the influence of an octahedral field can be found in Table 6.3. These results are given together in Table 6.4, which summarizes what we would expect to happen in a weak octahedral field.

Table 6.4 Terms Arising from d^2 in a Weak Octahedral Field

Free Ion	Octahedral Field
^1S	$^1A_{1g}$
^1G	$^1A_{1g}$, $^1T_{2g}$, ^1E_g, $^1T_{1g}$
^3P	$^3T_{1g}$
^1D	^1E_g, $^1T_{2g}$
^3F	$^3A_{2g}$, $^3T_{2g}$, $^3T_{1g}$

At the other extreme, that of an infinitely strong field, the lowest configuration of the ion will be t_{2g}^2, with higher configurations $t_{2g}e_g$ and e_g^2. These configurations will themselves give rise to terms, which will be split by interelectron repulsions if the field is less than infinitely strong, and it is these terms which must be correlated with the terms in the right-hand column of Table 6.4. We can work out the symmetry types of these terms by forming the direct products t_{2g}^2, $t_{2g}e_g$, and e_g^2 and then reducing the resulting representations by way of equations (4.65) and (4.43). The results of doing this are given in Table 6.5.

Table 6.5 Terms Arising from d^2 in a Strong Octahedral Field

Configuration	Terms
e_g^2	A_{1g}, E_g, A_{2g}
$t_{2g} \cdot e_g$	T_{1g}, T_{2g}
t_{2g}^2	A_{1g}, E_g, T_{1g}, T_{2g}

We are now able to begin to construct the correlation diagram, as shown in Fig. 16. The g subscript which occurs throughout has been omitted for simplicity.

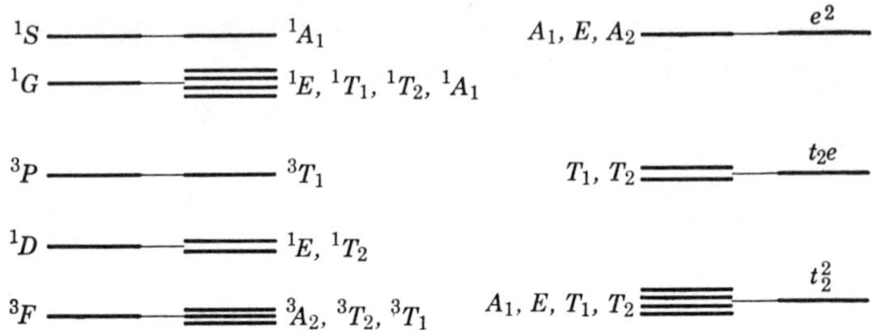

^1S ————— ————— $^1A_1$ A_1, E, A_2 ————— e^2

^1G ————— ≡≡≡ $^1E, {}^1T_1, {}^1T_2, {}^1A_1$

^3P ————— ————— $^3T_1$ T_1, T_2 ≡≡≡ t_2e

^1D ————— ≡≡≡ $^1E, {}^1T_2$

^3F ————— ≡≡≡ $^3A_2, {}^3T_2, {}^3T_1$ A_1, E, T_1, T_2 ≡≡≡ t_2^2

Fig. 16 Beginning of correlation diagram for the d^2 configuration in a field of symmetry O_h.

To proceed further we must make use of two important semi-empirical rules. These are

1. *There must be a* 1:1 *correspondence between the states at the two sides of the diagram.*
2. *States of the same multiplicity and symmetry type cannot cross (noncrossing rule).*

The only term of A_2 symmetry on the left-hand side of Fig. 16 is $^3A_2$, arising from ^3F, and this must correlate with the term arising from e^2. Both A_1 terms are singlets and the noncrossing rule leaves us no choice as to how they are correlated. The same applies to the ^1E terms.

Next we observe that the configuration t_2e must give rise to both a singlet and a triplet term of any given symmetry type, depending on whether the electrons in the two orbitals have their spins parallel or antiparallel. This observation enables us to fix the other correlation lines, and we note that the ground state is represented by a $^3T_1$ term on both sides of the diagram. The final correlation diagram is shown in Fig. 17. The curvature of the correlation lines is intended as a qualitative indication of the fact that levels of the same type tend to repel one another.

The d^2 correlation diagram for a tetrahedral arrangement of ligands may be constructed in a similar manner. It can be obtained directly from the diagram for the octahedral case by noting that the order of the configurations at the right-hand side of the diagram is reversed when the field is changed from octahedral to tetrahedral, but the multiplicity of the terms is unaffected because of the lack of

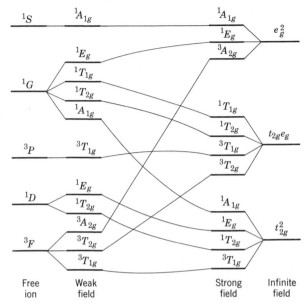

Fig. 17   Correlation diagram for the d^2 configuration in an octahedral field.

interaction between the electron spins and the ligand field.   The resulting correlation diagram is shown in Fig. 18.

The construction of correlation diagrams for d^n configurations by the foregoing method becomes rather difficult when the number of electrons is increased, mainly because of the difficulty of deciding the multiplicities of strong field terms.   For octahedral and tetrahedral fields we are aided by the fact that one diagram is the inverse of the other, as in Figs. 17 and 18.   In addition to this we note that a d^{10-n} configuration must lead to the same terms as d^n at the right-hand side of the diagram except that the order of the configurations, t_2^2, $t_2 e$, and e^2 in the case we have been considering, is the reverse of that for d^n.   We have already seen in section 6b that d^n and d^{10-n} give rise to the same terms at the left-hand side of the diagram. Hence the correlation diagrams for d^n (octahedral) and d^{10-n} (tetrahedral) are identical and are the inverse of the diagram for d^{10-n} (octahedral) and d^n (tetrahedral).

There is in addition a method due to Bethe, known as "descent of symmetry," which enables us to make unambiguous decisions as

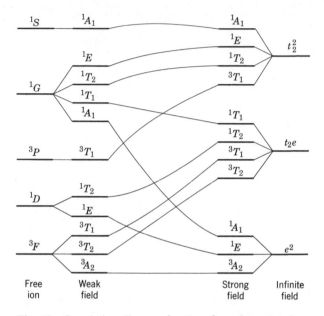

Fig. 18   Correlation diagram for the d^2 configuration in a tetrahedral field.

to the multiplicity of strong field terms.   To use this method we first need to know how the different representations of the point group O_h transform when the number of symmetry elements in the group is successively decreased and the point group is thereby changed to one of the subgroups of O_h, such as D_{4h} or C_{3v}.   The method of obtaining this information is very simple: as successive symmetry elements are eliminated the characters of the remaining ones stay the same as in the original point group.   These characters in general belong to a reducible representation of the new point group, and we say that in the environment of lower symmetry the original representation splits into the components of this reducible representation.   The most commonly useful results for O_h and its subgroups are given in Table 6.6.[2]

[2] More complete correlation tables are given in Wilson, Decius, and Cross, *Molecular Vibrations*, New York: McGraw-Hill, 1955.

Table 6.6   Correlation Table for the Point Group O_h

O_h	O	T_d	D_{4h}	D_{2d}	C_{4v}	C_{2v}	D_{3d}	C_{2h}
A_{1g}	A_1	A_1	A_{1g}	A_1	A_1	A_1	A_{1g}	A_g
A_{2g}	A_2	A_2	B_{1g}	B_1	B_1	A_2	A_{2g}	B_g
E_g	E	E	$A_{1g}+B_{1g}$	A_1+B_1	A_1+B_1	A_1+A_2	E_g	A_g+B_g
T_{1g}	T_1	T_1	$A_{2g}+E_g$	A_2+E	A_2+E	$A_2+B_1+B_2$	$A_{2g}+E_g$	A_g+2B_g
T_{2g}	T_2	T_2	$B_{2g}+E_g$	B_2+E	B_2+E	$A_1+B_1+B_2$	$A_{1g}+E_g$	$2A_g+B_g$
A_{1u}	A_1	A_1	A_{1u}	B_1	A_2	A_2	A_{1u}	A_u
A_{2u}	A_2	A_2	B_{1u}	A_1	B_2	A_1	A_{2u}	B_u
E_u	E	E	$A_{1u}+B_{1u}$	A_1+B_1	A_2+B_2	A_1+A_2	E_u	A_u+B_u
T_{1u}	T_1	T_2	$A_{2u}+E_u$	B_2+E	A_1+E	$A_1+B_1+B_2$	$A_{2u}+E_u$	A_u+2B_u
T_{2u}	T_2	T_1	$B_{2u}+E_u$	A_2+E	B_1+E	$A_2+B_1+B_2$	$A_{1u}+E_u$	$2A_u+B_u$

As an example of the method of descent of symmetry we consider once more the strong field terms for d^2 in an octahedral complex. We imagine the complex to undergo an infinitesimal distortion in such a way that the point group changes to one in which the degenerate representations E_g, T_{1g}, and T_{2g} reduce to sums of one-dimensional representations, and that a *different* set of one-dimensional representations results from each of the strong field terms of Table 6.5. With one-dimensional representations we can immediately decide the multiplicities, since two electrons in a single orbital can only produce a singlet state, while two electrons in different orbitals yield both a singlet and a triplet state.   Then when we return to point group O_h the multiplicities must remain the same.

For the present case we consider the change to be from O_h to C_{2v}. According to Table 6.6 we find that t_{2g} splits up into $a_1 + b_1 + b_2$, and e_g into $a_1 + a_2$.   Hence the energy level diagram is as shown in Fig. 19.

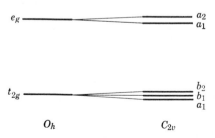

Fig. 19   Descent of symmetry.

From the configuration e_g^2 we obtain singlet a_1^2 and a_2^2 and singlet and triplet $a_1 a_2$. When the direct products are reduced this gives the terms $^1A_1$, $^1A_1$, $^1A_2$, and $^3A_2$. The terms which arise from e_g^2 in O_h are A_{1g}, A_{2g}, and E_g, which become A_1, A_2, and $A_1 + A_2$, respectively, when we transform to C_{2v}, and it follows that the strong field terms in O_h must be $^1A_{1g}$, ^1E_g and, $^3A_{2g}$.

We need not go through this procedure for the $t_{2g}e_g$ configuration because the electrons are already in different energy levels, and so we know that there must be equal numbers of singlet and triplet states. It would be worthwhile for the reader to carry this through as an exercise in using the method.

From t_{2g}^2 we obtain singlet a_1^2, b_1^2, and b_2^2 states and singlet and triplet $a_1 b_1$, $a_1 b_2$, and $b_1 b_2$ states. When the direct products are reduced the terms are $^1A_1$, $^1A_1$, $^1A_2$; $^1A_1$, $^1B_1$, $^1B_2$; $^3A_2$, $^3B_1$, $^3B_2$; and reference to Table 6.6 shows that the strong field terms in the octahedral environment must be $^1A_{1g}$, ^1E_g, $^1T_{2g}$, and $^3T_{1g}$, as we found previously.

Further correlation diagrams are given in the literature.[3] We shall defer consideration of the relationship between these diagrams and the spectra of complexes until Chapter 7. This chapter will conclude with a brief discussion of the magnetic properties of complexes in relation to ligand field theory.

6e   THE MAGNETIC PROPERTIES OF COMPLEXES

There are two main approaches to the investigation of the magnetic properties of complexes. The first is through measurements of static magnetic susceptibility, and the second is by way of paramagnetic resonance (or electron-spin resonance). We shall not consider either of these in very great detail.

Measurements of static magnetic susceptibility can yield valuable information about the number of unpaired electrons in a complex. When a substance is placed in a magnetic field of intensity **H** oersted the flux **B** in the substance is given by

$$\mathbf{B} = \mathbf{H} + 4\pi\mathbf{I} \tag{6.21}$$

---

[3] See for example, Tanabe and Sugano, *J. Phys. Soc. Japan*, **9**, 753 (1954); also the texts by Orgel and Ballhausen noted in Appendix I.

where **I** is the intensity of magnetization. The magnetic susceptibility of the substance per unit volume is then

$$\kappa = \mathbf{I}/\mathbf{H} \qquad (6.22)$$

The susceptibility per gram is χ and per gram-mole is χ_M, where

$$\chi_M = M\kappa/\rho \qquad (6.23)$$

Here M is the molecular weight and ρ is the density. The permeability μ of the substance is equal to $\mathbf{B}/\mathbf{H}$, so that

$$\mu = 1 + 4\pi\kappa \qquad (6.24)$$

The variation of the susceptibility with the absolute temperature may be represented empirically by the Curie Law

$$\chi = C/T \qquad (6.25)$$

or by the Curie-Weiss Law

$$\chi = C/(T - \theta) \qquad (6.26)$$

where C and θ are constants. For diamagnetic materials (i.e., for most substances) the susceptibility χ is small and negative; for paramagnetic substances (essentially those which contain unpaired electrons) it is somewhat larger and positive; while for a special class of materials, termed ferromagnetic, it is very large and positive. We are here interested only in the case of paramagnetism.

The effective magnetic moment of a molecule of a substance, μ_{eff}, is related to the molar susceptibility by the Langevin formula

$$\chi_M = N \cdot \frac{\mu_{\text{eff}}^2}{3kT} \qquad (6.27)$$

where N is Avogadro's number, k the Boltzmann constant, and μ_{eff} is expressed in Bohr magnetons.

According to Van Vleck[4]

$$\chi_M = N \cdot \frac{\beta^2 g^2 S(S + 1)}{3kT} \qquad (6.28)$$

where β is the Bohr magneton ($=e\hbar/2mc$), S is the resultant spin

[4] Van Vleck, *Theory of Electric and Magnetic Susceptibility*, London: Oxford University Press, 1932.

angular momentum of the molecule and the quantity g, known as the Landé g-factor, is given by

$$g = 1 + \frac{J(J + 1) - L(L + 1) + S(S + 1)}{2J(J + 1)} \qquad (6.29)$$

for an atom in Russell-Saunders coupling. For a free electron g is equal to 2.0023, the small departure from 2.0000 being due to relativistic effects.

From (6.27) and (6.28) we have

$$\mu_{\text{eff}}^2 = \beta^2 g^2 S(S + 1) \qquad (6.30)$$

where g may be put equal to 2.0 if the orbital contribution is small. This equation gives the "spin only" value of the magnetic moment, and for ions of metals of the first transition series, where spin-orbit coupling is not very important, it is a good approximation to the actual value. It is, however, seriously in error for elements of the later transition series.

The preceding equations enable us to deduce the total spin, S, of a complex from measurements of magnetic susceptibility. For an octahedral complex with between four and seven d electrons it is found that the total spin can have either of two values, depending on the extent of pairing of the electrons. This observation is readily accounted for in terms of ligand field theory, and the two possibilities for an octahedral complex of d^4 configuration are illustrated in Fig. 20.

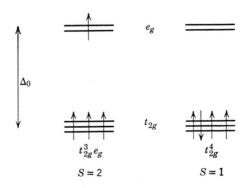

Fig. 20   High-spin and low-spin configurations for a d^4 ion in an octahedral field.

We note that the configuration t_{2g}^4 is lower in energy than $t_{2g}^3 e_g$ by an amount Δ_0; however, the energy of this configuration is increased by the repulsion energy of the electrons which are paired in one of the t_{2g} orbitals. Hence if this pairing energy exceeds Δ_0, the most stable configuration is $t_{2g}^3 e_g$, the total spin is 2 (in units of $\hbar$), and the ground state is a quintet. If the pairing energy is less than Δ_0, the total spin is 1, and the ground state is a triplet. It is apparent that a weak crystal field favors the formation of "high spin" complexes, while a strong field favors the formation of "low spin" complexes. It might be expected that similar effects would occur with tetrahedral complexes, but in this case only high spin complexes have been found, presumably because with $\Delta_t = 4\Delta_0/9$ the ligand field has to be twice as strong for Δ_t to exceed the pairing energy of the electrons.[5] The pairing energy can be estimated from spectroscopic data for the free ion. For d^4 to d^7 configurations the nature of the ground state in an octahedral complex changes at the point where Δ_0 is equal to the pairing energy, and this crossing is a feature of the correlation diagrams. For these configurations the multiplicities of the lowest free ion terms are always greater than the multiplicities of the lowest strong field terms (cf. Exercise 6.5).

The paramagnetic resonance approach[6] may be summarized briefly as follows:

The Hamiltonian operator for an atom in a magnetic field contains a term

$$\hat{H}'' = \beta \mathbf{H}(\mathbf{L} + 2\mathbf{S}) \tag{6.31}$$

where the symbols have the same significance as before. This additional perturbation removes the degeneracy associated with different values of M_J for a single level, and the energy separation of the levels, due to the Zeeman effect, is given by

$$\Delta E = g\beta \mathbf{H} \tag{6.32}$$

---

[5] The diamagnetic ion $ReCl_4^-$, in which rhenium has the d^4 configuration, has been suggested as a possible example of a tetrahedral, low spin complex. Recent work has shown, however, that the anion is a trimer, $Re_3Cl_{12}^{3-}$, of unusual stereochemistry. (Robinson, Fergusson and Penfold, *Proc. Chem. Soc. London*, p. 116, 1963.)

[6] See, for example, G. E. Pake, *Paramagnetic Resonance*, New York: W. A. Benjamin, 1962.

Experimentally a sample of material is held in a uniform magnetic field, often at liquid nitrogen temperature or below in order to drive most of the atoms into their lowest energy states, and then the sample is irradiated with microwaves of a fixed frequency ν. The strength of the magnetic field is then varied to locate the values of H for which $\Delta E = h\nu$, that is, for which radiation is absorbed. This amounts, in effect, to a determination of the value of g.

In a complex, the g factor is no longer given by (6.29) but, for example, for a d^n ion in an octahedral complex it is

$$g = 2 - 8\lambda/10D_q \qquad (6.33)$$

Here λ is the effective spin-orbit coupling constant for the atom as a whole, i.e., the equation

$$\hat{H}_{LS} = \lambda \mathbf{L} \cdot \mathbf{S} \qquad (6.34)$$

describes the multiplet splitting in the free atom.

An example, quoted by Ballhausen, is the complex ion $Ni(H_2O)_6^{2+}$, for which g is found experimentally to be 2.25.   From atomic spectroscopy we have $\lambda = -324$ cm^{-1}, so that $10D_q$ is predicted to be 10,700 cm^{-1}.   The first broad maximum in the absorption spectrum of this complex is definitely known to correspond to a transition of energy $10D_q$, and this occurs at 8,500 cm^{-1}.   The interpretation of the discrepancy is that λ is reduced from the free ion value, just as the spacings between atomic terms are reduced in the presence of the ligand field.

In complexes of less than cubic symmetry the g factor is anisotropic, i.e., it depends on the direction in which the magnetic field is applied.   This anisotropy can arise in complexes which are nominally of point group O_h or T_d as a result of "Jahn-Teller distortion." According to the Jahn-Teller theorem[7] the ground state of a system may not possess any orbital degeneracy, and all possible spin degeneracies must also be removed.   Thus if the ground state of a complex is expected to be of overall E or T symmetry, a distortion occurs in such a way that the orbital degeneracy is removed.   It appears that the effects of this distortion are very often too small to be detected (footnote 7), but there are a few cases where its occurrence is fairly well established.

[7] For a lucid discussion of the experimental consequences of this theorem see N. S. Ham, *Spectrochimica Acta*, **18**, 775 (1962).

A large amount of experimental and theoretical work is in progress in this field and for further details the reader is referred to Chapter 6 of Ballhausen's book (see Appendix I).

EXERCISES

6.1 (*a*) Work out the spectroscopic terms in Russell-Saunders coupling which arise from the following configurations: p^2, sp^2, d^2, sd^2, pd^2. Attempt to predict the order of the terms in each case.

(*b*) Write out the multiplet components of each of the terms in part (*a*).

6.2 Compare the energy differences between some corresponding multiplet components of chlorine, bromine, and iodine atoms. Relate your observations to the relative appropriateness of Russell-Saunders coupling schemes for these atoms. (Find the energy levels in Circular 467 of the National Bureau of Standards.)

6.3 Verify a few of the entries in Table 6.3.

6.4 (*a*) Show that in a field of symmetry D_{4h} a set of d orbitals transforms as follows: d_{z^2}, A_{1g}; $d_{x^2-y^2}$, B_{1g}; d_{xy}, B_{2g}; d_{yz}, d_{xz}, E_g.

(*b*) Work out the order of the energy levels for a square planar MX_4 complex in the strong-field case. (*Note*: The four ligands occupy the corners of a square in the xy plane.)

6.5 Show that for octahedral complexes of atoms of configurations $d^4 - d^7$ the numbers of unpaired electrons are as given in the following table:

Configuration	Low Spin Complexes	High Spin Complexes
d^4	2	4
d^5	1	5
d^6	0	4
d^7	1	3

6.6 It is possible to define a "spectrochemical series" of ligands, such that the characteristic visible absorption band of a complex of a given metal ion, with a particular symmetry, moves to a shorter wavelength when one ligand is replaced by another which is higher in the series. Such a series is $I^- < Br^- < Cl^- < OH^- < F^- < H_2O <$ pyridine $< NH_3 <$ ethylenediamine $< CN^-$. Interpret this order in terms of relative Δ values, and suggest how it might be related to the occurrence of high spin or low spin complexes of a given metal. Note that this is not quite the same series as that which corresponds to the order of the stability constants of complexes. (Ballhausen, *op. cit.*, p. 221).

7

Topics Related to Spectroscopy

7a INTRODUCTION

The subject of spectroscopy is of great importance to quantum mechanics because it provides the most direct experimental evidence for the existence of discrete energy levels in atomic and molecular systems. One consequence of this is that many of the most striking developments of the theory have been stimulated by the need to account for spectroscopic observations; examples which readily come to mind are the Pauli exclusion principle and the hypothesis of electron spin.

The field of atomic and molecular spectroscopy is obviously too large to be covered adequately in a single volume, and we are therefore faced with the problem of how to achieve something worthwhile in the space of one chapter. Our solution will be to concentrate on a few selected topics, notably radiation theory, electronic states of molecules, molecular vibrations, and to leave other important matters, such as the reasons for the characteristic appearances of different types of band spectra, to be dealt with in the texts which are listed in Appendix I. The topics for discussion have been selected mainly on the grounds that they are likely to be of interest to persons who are not purely spectroscopists but who wish to make use of spectroscopy in other fields, for example, photochemistry, chemical kinetics, or statistical thermodynamics. Another criterion has been that these topics should constitute logical extensions of the methods and ideas of earlier chapters. Thus they will serve to illustrate, among other things, the use of time-dependent perturbation theory and the application of group theory to molecular dynamics. This chapter will conclude with a discussion of some representative spectra of polyatomic molecules.

7b RADIATION THEORY

This section contains an account of radiation theory in the form in which it is most relevant to spectroscopy. The approach is by way of the time-dependent perturbation theory which was introduced in Chapter 3. In the presence of electromagnetic radiation there is a small, time-dependent perturbation of the Hamiltonian operator of a system as a result of interaction between the oscillating electromagnetic field and the system's instantaneous dipole moment.[1] This perturbation causes the system to switch back and forth repeatedly from one state to another, with absorption or emission of radiation whose frequency is such that $h\nu$ (or $\hbar\omega$) is equal to the energy difference between the states. The rate at which the system undergoes these induced radiative transitions is proportional to the energy density of the appropriate radiation, and we shall derive an expression for the proportionality factor. A simple thermodynamic argument will then enable us to deduce the probability factor for spontaneous emission from an excited state of a system. These results will later enable us to determine *selection rules* for various types of spectroscopic transitions.

We consider a system for which an arbitrary wave function is

$$\Psi = a_1\Psi_1 + a_2\Psi_2 + \cdots a_r\Psi_r \tag{7.1}$$

The functions Ψ_n are assumed to be of the form

$$\Psi_n(q, t) = \phi_n(q) \cdot e^{-iE_n t/\hbar} \tag{7.2}$$

so that

$$\hat{H}_0\Psi_n = E_n\Psi_n \tag{7.3}$$

where $\hat{H}_0$ is the unperturbed Hamiltonian operator for the system. If the actual Hamiltonian is

$$\hat{H} = \hat{H}_0 + \hat{H}' \tag{7.4}$$

---

[1] Interaction with the system's *induced* dipole moment gives rise to the Raman effect, which is therefore dependent on the polarizability of the system. (Cf. Eyring, Walter, and Kimball, *Quantum Chemistry*, New York: Wiley, 1944, p. 121.)

where $\hat{H}'$ is a time-dependent perturbation, then it was shown in Chapter 3 that the coefficients in the expansion (7.1) vary with time at a rate which is given by

$$\frac{da_m}{dt} = \frac{-i}{\hbar} \sum_n a_n H'_{mn}(t) \tag{7.5}$$

where $H'_{mn}(t)$ is the matrix element of $\hat{H}'$ with respect to $\Psi_m(q, t)$ and $\Psi_n(q, t)$.

In terms of the matrix elements of $\hat{H}'$ with respect to the stationary state functions $\phi_m(q)$ and $\phi_n(q)$ equation (7.5) becomes

$$\frac{da_m}{dt} = \frac{-i}{\hbar} \sum_n a_n \cdot H'_{mn} \cdot e^{i\omega_{mn}t} \tag{7.6}$$

where

$$\omega_{mn} = \frac{(E_m - E_n)}{\hbar} \tag{7.7}$$

Now let us take as the perturbation a light wave, moving along the z axis, whose electric vector along the x axis is

$$E_x = 2E_x^0 \cos(\omega t - 2\pi z/\lambda) \tag{7.8}$$

We then have

$$\hat{H}' = \mu^x \cdot E_x \tag{7.9}$$

where μ^x is the x component of the dipole moment of the system.[2]

The dimensions of molecules can usually be assumed to be small compared with λ, so that we may take E_x to be constant over the molecule and neglect the term $2\pi z/\lambda$ in equation (7.8).

Hence we obtain

$$\frac{da_m}{dt} = \frac{-2E_x^0 i}{\hbar} \sum_n a_n \mu_{mn}^x \cos \omega t \cdot e^{i\omega_{mn}t} \tag{7.10}$$

where μ_{mn}^x is a matrix element of the x component of the dipole moment.

---

[2] Here we are neglecting the interaction between the electrons (whose displacement is responsible for the appearance of a dipole moment) and the magnetic vector of the light wave. This is smaller than the interaction with the electric vector in the ratio v/c, where v is the velocity of the electron and c the velocity of light. For an electron in a hydrogen atom v/c is equal to $1/137$, and in the present state of the experimental field an error of this magnitude is quite unimportant.

It is useful to express $\cos \omega t$ in the form $\frac{1}{2}(e^{i\omega t} + e^{-i\omega t})$, when (7.10) becomes

$$\frac{da_m}{dt} = \frac{-E_x^0 i}{\hbar} \sum_n a_n \mu_{mn}^x [e^{i(\omega_{mn} + \omega)t} + e^{i(\omega_{mn} - \omega)t}] \qquad (7.11)$$

Let us now simplify the physical situation by supposing that the system is initially in the state described by Ψ_n, when $a_n = 1$ and all the other coefficients a_m are zero. The result is

$$\frac{da_m}{dt} = \frac{-E_x^0 i}{\hbar} \cdot \mu_{mn}^x [e^{i(\omega_{mn} + \omega)t} + e^{i(\omega_{mn} - \omega)t}] \qquad (7.12)$$

If we assume that the probability of the system undergoing a transition is small, so that a_n remains close to unity, we can integrate (7.12) with respect to time and so obtain the value of a_m at a time t. Using the boundary condition $a_m = 0$ at $t = 0$, we obtain the equation

$$a_m = \frac{E_x^0 \cdot \mu_{mn}^x}{\hbar} \left[\frac{1 - e^{i(\omega_{mn} + \omega)t}}{\omega_{mn} + \omega} + \frac{1 - e^{i(\omega_{mn} - \omega)t}}{\omega_{mn} - \omega} \right] \qquad (7.13)$$

Now if ω_{mn} is positive, that is, $E_m > E_n$, the term with $(\omega_{mn} - \omega)$ in the denominator will become very large when ω_{mn} and ω are approximately equal. This corresponds to the situation where a quantum is absorbed when the system undergoes a transition from Ψ_n to Ψ_m. The term with $(\omega_{mn} + \omega)$ in the denominator, on the other hand, remains small because both ω and ω_{mn} are typically very large (about 10^{14} sec.$^{-1}$ for visible light). Hence it is a good approximation to write

$$a_m = \frac{E_x^0 \cdot \mu_{mn}^x}{\hbar} \cdot \frac{1 - e^{i(\omega_{mn} - \omega)t}}{\omega_{mn} - \omega} \qquad (7.14)$$

The probability of finding the system in the state Ψ_m after a time t is given by $a_m^* \cdot a_m$, that is, by

$$|a_m|^2 = \frac{2(E_x^0)^2 \cdot |\mu_{mn}^x|^2}{\hbar^2} \cdot \frac{1 - \cos(\omega_{mn} - \omega)t}{(\omega_{mn} - \omega)^2} \qquad (7.15)$$

$$= \frac{(E_x^0)^2 \cdot |\mu_{mn}^x|^2}{\hbar^2 \cdot t^{-2}} \cdot \frac{\sin^2 [(\omega_{mn} - \omega)/2]t}{[(\omega_{mn} - \omega)/2]^2 t^2} \qquad (7.16)$$

This result is not complete, however, until we integrate over all radiation frequencies. The probability of absorption of a quantum is extremely small except over a small frequency range near ω_{mn}, and so we can treat E_x^0 as a constant during the integration, putting it equal to $E_x^0(\nu_{mn})$. Also, for this reason, we can integrate from $-\infty$ to ∞, and so avoid complications with the integration limit $\omega = 0$. We make use of the definite integral

$$\int_{-\infty}^{\infty} \frac{\sin^2 y \, dy}{y^2} = \pi \qquad (7.17)$$

and carry out the integration with respect to $\nu = \omega/2\pi$. The result is

$$|a_m|^2 = [E_x^0(\nu_{mn})]^2 \cdot |\mu_{mn}^x|^2 \cdot t/\hbar^2 \qquad (7.18)$$

where $E_x^0(\nu_{mn})$ is the electric vector of the radiation of frequency $\omega_{mn}/2\pi = \nu_{mn}$.

The rate of transition from state n to state m is then

$$d|a_m|^2/dt = [E_x^0(\nu_{mn})]^2 \cdot |\mu_{mn}^x|^2/\hbar^2 \qquad (7.19)$$

In radiation theory the rate at which this transition occurs is usually expressed in terms of the Einstein coefficient for induced absorption of radiation, B_{nm}, where

$$d|a_m|^2/dt = B_{nm} \cdot \rho(\nu_{mn}) \qquad (7.20)$$

Here $\rho(\nu_{mn}) \cdot d\nu$ is the radiation energy density, measured in ergs per cubic centimeter, in the frequency range between ν_{mn} and $\nu_{mn} + d\nu$.

For isotropic radiation it can be shown that

$$\rho(\nu) = \frac{1}{4\pi} ([\bar{E}_x(\nu)]^2 + [\bar{E}_y(\nu)]^2 + [\bar{E}_z(\nu)]^2)$$

$$= \frac{3}{4\pi} \cdot [\bar{E}_x(\nu)]^2$$

$$= \frac{3}{2\pi} \cdot [E_x^0(\nu)]^2 \qquad \text{(Cf. equation 7.8.)} \qquad (7.21)$$

where a bar indicates a mean value and an extra factor of 2 arises from the fact that the mean value of $\cos^2 2\pi\nu t$ is $\frac{1}{2}$. Hence we obtain, from (7.19), (7.20), and (7.21),

$$B_{nm} = \frac{2\pi}{3\hbar^2} (|\mu_{mn}^x|^2 + |\mu_{mn}^y|^2 + |\mu_{mn}^z|^2)$$

$$= \frac{2\pi}{3\hbar^2} |\mu_{mn}|^2 \qquad (7.22)$$

where we have included the additional transition probabilities for *y*- and *z*-polarized components of the radiation. Equation (7.22) is the basis of spectroscopic selection rules for dipole radiation. The quantity μ_{mn} is usually called the "transition moment."

If we had chosen the first exponential term of equation (7.13) as the one responsible for the transition, we would have obtained an identical expression for the Einstein transition probability for *stimulated emission*, B_{mn}. Hence we have the result

$$B_{nm} = B_{mn} \tag{7.23}$$

For a complete discussion of absorption and emission of radiation we also require the Einstein transition probability, A_{mn}, for *spontaneous emission* from an excited state. A direct calculation of this quantity would involve us in quantum electrodynamics, but fortunately the value of A_{mn} can be deduced from that for B_{mn} with the aid of a simple thermodynamic argument, due to Einstein, which is as follows:

We consider a system which contains a large number of molecules in equilibrium with black-body radiation. The relative populations of states Ψ_m and Ψ_n are given by

$$N_m/N_n = e^{-E_m/kT}/e^{-E_n/kT}$$
$$= e^{-h\nu_{mn}/kT} \tag{7.24}$$

The rate at which transitions from state *n* to state *m* occur is $N_n \cdot B_{nm} \cdot \rho(\nu_{mn})$, while the rate at which transitions occur in the reverse direction is equal to $N_m(B_{mn} \cdot \rho(\nu_{mn}) + A_{mn})$. At equilibrium these two rates are equal, so we must have

$$e^{-h\nu_{mn}/kT} = B_{nm} \cdot \rho(\nu_{mn})/(B_{mn} \cdot \rho(\nu_{mn}) + A_{mn}) \tag{7.25}$$

or

$$\rho(\nu_{mn}) = \frac{A_{mn}/B_{nm}}{e^{h\nu_{mn}/kT} - 1} \tag{7.26}$$

But according to the Planck radiation law

$$\rho(\nu_{mn}) = \frac{8\pi h\nu_{mn}^3}{c^3} \cdot \frac{1}{e^{h\nu_{mn}/kT} - 1} \tag{7.27}$$

from which it follows that

$$A_{mn} = \frac{8\pi h\nu_{mn}^3}{c^3} \cdot B_{nm} \tag{7.28}$$

$$= \frac{32\pi^3 \nu_{mn}^3}{3c^3 \hbar} |\mu_{mn}|^2 \tag{7.29}$$

For allowed transitions, i.e., those which are not forbidden by selection rules, the value of A_{mn} is typically 10^6 to 10^9 sec.$^{-1}$, and in the absence of other deactivation processes it is equal to the reciprocal of the mean lifetime of the excited state. Stimulated emission is generally much less important in practice than spontaneous emission, however, this is not true of lasers,[3] where it is arranged that a large fraction of the atoms in a crystal or a gas discharge tube are in excited states. In this situation a photon which is emitted by one atom has a good chance of stimulating the emission of an identical photon from another atom nearby. Laser action can only occur if the probability that a photon will stimulate further emission is greater than the probability that it will be absorbed.

The Einstein coefficient for stimulated absorption can easily be related to the experimental extinction coefficient $\varepsilon(\nu)$ of a substance (cf. Exercise 7.2). The result is

$$\int \varepsilon(\nu) \, d\nu = \frac{2\pi N h}{2.303 \times 1000c} \cdot \nu_{mn} \cdot B_{nm} \qquad (7.30)$$

where N is Avogadro's number, c is the velocity of light in the absorbing medium, and the extinction coefficient $\varepsilon(\nu)$ is defined by

$$\varepsilon(\nu) = \frac{1}{\mathscr{C} \cdot l} \log_{10} \frac{I_0}{I} \qquad (7.31)$$

Here $\mathscr{C}$ is the concentration of the absorber in moles per litre, l is the length of the light path in the medium and I_0 and I are the intensities of the incident and emergent light beams, respectively. The integration over $d\nu$ is to take account of the fact that absorption does not occur at one single frequency but over an absorption line, or even a band, of finite width.

Equation (7.30) relates the area under an absorption line or band to the Einstein coefficient B_{nm}. From B_{nm} we can calculate A_{mn} with the aid of equation (7.28), and hence obtain an estimate of the lifetime of the excited state. Lifetimes estimated in this way usually agree to within about $\pm 20\%$ with directly measured values for the excited states of molecules.[4]

As indicated before, the main use of the results of this section is in the determination of selection rules for various types of spectro-

[3] Lengyel, *Lasers*, New York: Wiley, 1962.
[4] See, for example, Strickler and Berg, *J. Chem. Phys.*, 37, 814 (1962).

scopic transitions. We shall give detailed consideration to selection rules in section 7d when we know more about the excited states of molecules.

If the transition moment μ_{mn} is zero, the transition is obviously forbidden for electric dipole radiation. Such a transition may still take place with emission or absorption of magnetic dipole or electric quadrupole radiation. The magnetic dipole and electric quadrupole transition probabilities arise when the variation of the radiation field over the dimensions of the molecule is not neglected. They are generally much smaller than the transition probability for electric dipole radiation; if the results are expressed in terms of a quantity f, known as the oscillator strength, we find typically that for allowed transitions

$$f(\text{electric dipole}) \cong 1$$
$$f(\text{magnetic dipole}) \cong 10^{-5} \qquad (7.32)$$
$$f(\text{electric quadrupole}) \cong 10^{-7}$$

The oscillator strength is defined as unity for an oscillator which consists of a single electron bound to a center of attraction by Hooke's law forces, so that it undergoes simple harmonic motion about its equilibrium position. (The odd valence electron of a sodium atom is a fair approximation to this.) For such a hypothetical system we may calculate in turn the values of B_{nm} and the integrated extinction coefficient. The latter works out to be

$$\int \varepsilon(\bar{\nu}) \, d\bar{\nu} = 2.31 \times 10^8 \text{ cm.}^{-2} \text{ mole}^{-1} \text{ liter} \qquad (7.33)$$

where the integral is expressed in terms of wave numbers, $\bar{\nu}$, measured in cm.$^{-1}$, since these are more convenient in practice than the frequency ν (in sec.$^{-1}$). The f value of an actual transition is then given by the ratio of its integrated extinction coefficient to the value for $f = 1$. Experimental data on line or band intensities is usually quoted in the literature in terms of f values rather than Einstein coefficients.

7c ELECTRONIC STATES OF MOLECULES

In this section we shall consider the question of how the different types of electronic states of molecules may be distinguished and characterized. The classification of electronic states of diatomic molecules, and also of linear polyatomic molecules, is analogous to the classification of atomic states which was discussed in Chapter 6. For nonlinear polyatomic molecules, on the other hand, electronic states are best classified by their symmetry species, i.e., they acquire the label of the representation of the molecular point group according to which the overall electronic wave function transforms. These two systems of nomenclature can be made to overlap by treating wave functions of linear molecules as bases for representations of the point groups $D_{\infty h}$ or $C_{\infty v}$. Representations of these groups may thus be labeled either in accordance with the systematic procedure for point groups which was outlined in Chapter 4 or according to the equally systematic procedure which spectroscopists have developed for use with diatomic molecules. Both sets of labels for representations of $C_{\infty v}$ and $D_{\infty h}$ are given in Appendix II. We shall, however, deal separately with diatomic and polyatomic molecules, as is in keeping with both the historical development and the present situation in the field.

Diatomic Molecules. The pattern of energy levels in a series of isoelectronic diatomic molecules is generally quite similar to the pattern of levels for an atom with the same number of electrons.[5] Therefore as a first approximation we can consider a diatomic molecule as a "united atom," with the two nuclei very close together, and can attempt to classify the electronic states of the molecule in terms of the resultant orbital and spin angular momenta of the electrons, as was done for atoms in Chapter 6. This approach is facilitated by the fact that in a diatomic molecule there is an intense field along the internuclear axis, due to the presence of the two positively charged nuclei, and so it is to be expected that the various angular momenta will be quantized with respect to this axis.

In addition to the angular momentum associated with the electrons, a molecule can also possess angular momentum due to rotation of the molecule as a whole about an axis perpendicular to the internuclear axis (cf. the rigid rotator, Chapter 2). Therefore, since it is the *total* angular momentum of the system which is conserved, we must consider the manner in which this rotation may be combined with the orbital and spin angular momenta of the electrons. We

[5] See, for example, Walker and Straw, *Spectroscopy*, Vol. II, London: Chapman and Hall, 1962, p. 46.

shall not find it necessary to consider the effect of nuclear spin of the atoms comprising the molecule, because, although the resultant nuclear spin has a profound effect on the statistical weights of rotational energy levels, its effect on the electronic levels is generally negligible.

In a diatomic molecule the individual orbital angular momentum vectors **l** of the electrons combine to form a resultant **L**, whose component along the internuclear axis is labeled Λ. The magnitude of this component is $\Lambda\hbar$, while the magnitude of **L** is $\sqrt{L(L+1)}\cdot\hbar$. Electronic terms with $\Lambda = 0, 1, 2, 3, \cdots$ are labeled with the letters $\Sigma, \Pi, \Delta, \Phi, \cdots$ by analogy with the atomic $S, P, D, F, \cdots$ etc.

The individual electron spins similarly combine to form a resultant **S** whose component along the internuclear axis is labeled Σ (not to be confused with the symbol for a state for which $\Lambda = 0$). Then, as in Russell-Saunders coupling for an atom, Λ and Σ combine to form a resultant Ω which is analogous to the atomic **J**. The multiplicity of a term is $2S + 1$ and this is written as a superscript in front of the term symbol. In contrast to the situation with atomic states the full multiplicity is observed even for $S > \Lambda$. Gaydon[6] quotes the example of the $^4\Pi$ term of O_2^+, in which four sublevels appear corresponding to $\Omega = 2\frac{1}{2}, 1\frac{1}{2}, \frac{1}{2},$ and $-\frac{1}{2}$. The value of Ω for a particular state is written as a subscript after the term symbol, for example, with $\Lambda = 1$ and $S = 1$ we expect to find the states $^3\Pi_2, ^3\Pi_1,$ and $^3\Pi_0$. The order of energy levels is usually such that the energy increases with increasing Ω. When this is not the case the levels are said to be inverted, and this is usually indicated by a small subscript i placed after the term symbol. Thus the ground state of OH is usually written $^2\Pi_i$ because the $^2\Pi_{1\frac{1}{2}}$ level lies below $^2\Pi_{\frac{1}{2}}$.

In addition to these superscripts and subscripts the states of homonuclear diatomic molecules (C_2, N_2, O_2) are labeled g or u according to whether the total wave function is symmetrical or antisymmetrical with respect to inversion in the center of symmetry of the molecule. States for which $\Lambda = 0$, i.e., Σ states, are labeled Σ^+ or Σ^- according to the symmetry of the wave function with respect to reflection in a plane through the internuclear axis. This symmetry property is intimately connected with the symmetry of the molecular rotational levels, as will be explained shortly.

[6] A. G. Gaydon, *Dissociation Energies*, London: Chapman and Hall, 1953.

Spectroscopists often use letters in roman type to distinguish states of the same kind, it being conventional to use X for the ground state and A, B, C, $\cdots$ for successive excited states of the same multiplicity as the ground state. The only exception to this is N_2, where through long usage the letters A, B, and C are used for ${}^3\Sigma_u^+$, ${}^3\Pi_g$, and ${}^3\Pi_u$, respectively, although the ground state is $X({}^1\Sigma_g^+)$. The excited singlets in this case (${}^1\Pi_g$ and ${}^1\Pi_u$) are labeled a and b for short. Details of the states of a large number of diatomic molecules are listed by Herzberg.[7]

Next we consider the effect of rotation of the molecule as a whole. As mentioned in Chapter 2, a diatomic molecule is quite a good approximation to a rigid rotator. For the simplest case, a molecule in a ${}^1\Sigma$ state, the rotational energy levels are given by

$$E_r = B \cdot K(K + 1) \tag{7.34}$$

where it is customary to use K or J for the rotational quantum number and the rotational energy constant B is equal to $\hbar^2/2I$, I being the moment of inertia of the molecule. The quantity B has different values in different electronic states and varies slightly with the vibrational quantum number in any one state. In practice it is also necessary to include further terms in (7.34) to allow for centrifugal stretching of the molecule when K is large.

There are several different ways in which the various angular momentum vectors may be combined together (coupled) to form a resultant. The different possibilties are referred to as *Hund's coupling cases* and are as follows:

Case a: In this form of coupling Λ and Σ combine to form a resultant Ω which in turn combines with the molecular rotation to produce a resultant $\mathbf{J}$. J takes the values Ω, $\Omega + 1$, $\Omega + 2$, etc., and the rotational energy levels are given by

$$E_r = B(J(J + 1) - \Omega^2) \tag{7.35}$$

The selection rules for rotational transitions are determined by J, which is a "good" quantum number, i.e., the momentum to which it refers is strictly conserved. In this coupling case the multiplet splitting due to electron spin is fairly large. Case a coupling holds

[7] G. Herzberg, *Molecular Spectra and Molecular Structure*, Vol. I, *Spectra of Diatomic Molecules*, 2nd Edition, New Jersey: Van Nostrand, 1950, Table 39. See also P. G. Wilkinson, *J. Mol. Spec.*, **6**, 1 (1961).

for most molecular states other than singlets and Σ states, except when the molecule contains a very heavy atom or two very light atoms.

Case b: Here the spin is relatively unimportant. The molecular rotation combines with Λ to give a resultant $\mathbf{K}$, which then combines with the spin, if any, to produce a final resultant $\mathbf{J}$. In this case K is a good quantum number, taking the values Λ, $\Lambda + 1$, $\Lambda + 2$, etc., and the rotational energy levels are given by

$$E_r = B(K(K + 1) - \Lambda^2) + \text{small terms due to spin splitting}$$
(7.36)

This coupling case holds for all singlets and all Σ states except where very heavy atoms or highly excited states are involved. Molecules with two very light atoms (MgH and OH are examples quoted by Gaydon) also approximate to case b coupling.

Case c: This coupling case, which applies to molecules containing one or two very heavy atoms, is analogous to $j - j$ coupling for atoms. The $\mathbf{l}$ and $\mathbf{s}$ vectors of individual electrons are combined to form resultants $\mathbf{j}$ and these then combine to form a resultant $\mathbf{\Omega}$. The rotational energy levels are given by equation (7.35). It is possible to label the states as if they conformed to case a, but Λ and Σ are no longer good quantum numbers and the states are properly labeled with just the value of Ω.

Case d: When an electron is very highly excited it may be regarded as moving in an approximately Bohr-type orbit about the nuclei and inner electrons. The energy of the molecule is then largely determined by the principal quantum number of this electron, the effects of coupling $\mathbf{l}$ and $\mathbf{s}$ being relatively small. This case is approached for very highly excited states of diatomic molecules. The quantum numbers Λ and Σ are not defined, and states are usually labeled as if they belonged to atoms, for example, ^2P. The rotation of the molecule is given a quantum number R and the rotational energy levels are described by the equation

$$E_r = BR(R + 1) + \text{small terms which produce splitting first into}$$
$$2L + 1 \text{ components and then into } 2S + 1$$
$$\text{subcomponents.} \quad (7.37)$$

Case e: is not of practical importance.

Molecular states in case *a* or *b* for which $\Lambda > 0$ possess an inherent twofold degeneracy, since the orbital momentum **L** has components with Λ equal to $L, L - 1, L - 2, \cdots 1, 0, -1, \cdots -(L - 1), -L$, and the states with equal positive and negative values of Λ coincide. This is also true in case *c* with $\Omega > 0$. The interaction of $\boldsymbol{\Lambda}$ or $\boldsymbol{\Omega}$ with the molecular rotation leads to a removal of this degeneracy, each rotational level being split into two components. This effect is termed Λ-type doubling. One component then has $(+)$ symmetry and the other $(-)$ symmetry with respect to reflection of the rotational wave function in the origin (transformation of x, y, z to $-x$, $-y$, $-z$). The splitting is not more than a few cm.$^{-1}$, increasing with increasing J, and is usually unimportant except in the microwave (pure rotational) spectra of a few molecules, such as NO, whose ground states are not Σ states.

For Σ states, or 0 states in case *c*, this degeneracy is absent and all rotational levels are either of $(+)$ symmetry with K even and $(-)$ symmetry with K odd or vice-versa. States for which the rotational levels are $(+)$ with K even are labeled Σ^+ or 0^+, while those for which the rotational levels are $(-)$ with K even are labeled Σ^- or 0^-. It can be shown that this is equivalent to our previous statement that Σ^+ (and 0^+) states are those for which the wave function is symmetrical with respect to reflection in a plane through the internuclear axis. The symmetry properties of states are very important in connection with the spectroscopic selection rules, as will become apparent in section 7d.

An interesting topic, which we are not able to consider here because of lack of space, concerns the question of how to correlate the states of diatomic molecules with particular states of the separated atoms. Rules for determining these correlations have been derived by Wigner and Witmer and are discussed in Chapter 6 of Herzberg's *Spectra of Diatomic Molecules* (footnote 7) and in Chapter 3 of Gaydon's *Dissociation Energies* (footnote 6).

Polyatomic Molecules. The study of electronic states of polyatomic molecules has tended to progress in several directions at once as a result of effort in the fields of theoretical chemistry, photochemistry, spectroscopy, and fluorescence and phosphorescence. One consequence of this multilateral growth has been the coexistence, harmonious or otherwise, of a number of competing systems of nomenclature for molecular states and spectroscopic transitions. The corresponding confusion of symbols in the literature has probably served to keep each pocket of research moving steadfastly in its own particular direction, despite attempts which have been made to correlate the opposing systems and perhaps even eliminate one or two of them.[8]

In this account we shall concentrate on the nomenclature derived from molecular orbital theory, in which a state may be described either by its electron configuration, i.e., by specifying the number of electrons which are present in each type of bonding or antibonding molecular orbital, or by the symmetry species of the overall electronic wave function. It usually happens that an excited configuration gives rise to several states of different symmetry type, so that these descriptions are complementary rather than mutually exclusive. A complete description of a state also requires the specification of the spin multiplicity, $2S + 1$, and this is written as a superscript before the symbol for the symmetry species. We have in fact already met this system in Chapter 6 when we were discussing the energy levels of atoms in complexes. To show how it works in other cases we will consider two typical examples, namely benzene and form-aldehyde.

1. *Benzene* (D_{6h}): The ground state configuration is $a_{2u}^2 \cdot e_{1g}^4$, where we adhere to the custom of using lower case letters for the symmetry species of the orbitals. A "filled shell" configuration of this type gives rise to an overall wave function which is totally symmetric, $^1A_{1g}$ in this case, and with few exceptions the ground states of molecules always belong to totally symmetric representations.[9] The first excited configuration is $a_{2u}^2 \cdot e_{1g}^3 \cdot e_{2u}$ and this must give rise to equal numbers of singlet and triplet states of which, by Hund's rule, the triplet states will be of lower energy than the corresponding

[8] The various systems are described in several of the references to this chapter in Appendix I. The present situation is very well summarized by Pitts, Wilkinson and Hammond, "The Vocabulary of Photochemistry," *Advances in Photochemistry*, Vol. I, Ed. Noyes, Hammond and Pitts, New York: Interscience, 1963.

[9] It is obvious that a direct product of the type a^2 or b^2 can give rise only to a totally symmetric representation because all the characters become $+1$. A direct product of type e^2 or t^2 reduces into several component representations, and in this case it is not difficult to show, using descent-of-symmetry ideas (cf. section 6d), that configurations with all electrons paired must correlate with totally symmetric representations.

singlets. We can reduce the representation which corresponds to this direct product (it is only necessary to consider the product $e_{1g} \cdot e_{2u}$ since the rest is totally symmetric), and we then find that the excited configuration gives rise to states of symmetry species B_{1u}, B_{2u}, and E_{1u}. It is very difficult to decide the order of these states by means of a theoretical calculation because of the difficulty of calculating the interelectron repulsion energies which are primarily responsible for splitting the configuration into its separate terms. Nevertheless, several such calculations have been carried out with the result that the order of the singlet states is predicted to be $^1A_{1g}$, $^1B_{2u}$, $^1B_{1u}$, $^1E_{1u}$, and that the lowest triplet is $^3B_{1u}$. This order appears to be consistent with the observed absorption spectrum (see section 7f).

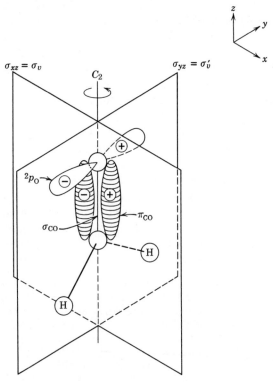

Fig. 21   Formaldehyde, symmetry of bonding orbitals.

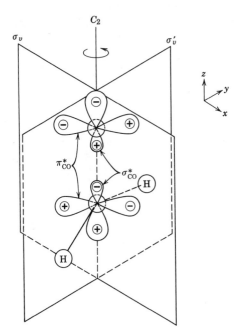

Fig. 22 Formaldehyde, symmetry of antibonding orbitals.

2. *Formaldehyde* (C_{2v}): Formaldehyde may be regarded as the
prototype of all compounds which contain a carbonyl group as a
chromophore. The point group of the molecule is C_{2v} and the
symmetry of the orbitals may be inferred from Fig. 21.

The total electronic wave function of the molecule can be
analyzed into the configuration $1s_C^2 \cdot 1s_O^2 \cdot 2s_O^2 \cdot \sigma_{CH}^2 \cdot \sigma_{CH'}^2 \cdot \sigma_{CO}^2 \cdot \pi_{CO}^2 \cdot 2p_O^2$.
Here the carbon atom uses sp^2 hybrid orbitals for its three σ bonds
while the oxygen atom uses practically pure unhybridized p orbitals.
Because all of the orbitals are doubly occupied, the wave function
of the ground state is of symmetry species A_1. The highest orbitals
in the configuration are $2p_O$ and π_{CO}, and it is to be expected that
the most important excited configurations will involve the promotion of electrons from these orbitals into antibonding π_{CO}^* or
possibly σ_{CO}^* orbitals. The symmetry of the antibonding orbitals
is illustrated in Fig. 22.

To determine the symmetry species of the states which arise from the excited configurations it is necessary to know the symmetry species of the individual orbitals.   These can be found by inspection from Figs. 21 and 22, with the aid of the C_{2v} character table (Appendix II).   The results are listed in Table 7.1.

Table 7.1

Orbital	Symmetry Species
σ_{CO}	a_1
σ_{CO}^*	a_1
π_{CO}	b_1
π_{CO}^*	b_1
$2p_O$	b_2

The spectroscopic terms which arise from the first few excited configurations of formaldehyde, as obtained by reducing the appropriate direct products and allowing for all possible spin multiplicities, are listed in Table 7.2.

Table 7.2

Configuration of HCHO	Resulting Electronic Terms	
(Inner electrons)$\cdot\pi^2\cdot 2p^2$	$^1A_1$	
(Inner electrons)$\cdot\pi^2\cdot 2p\cdot\pi^*$	$^1A_2$	$^3A_2$
(Inner electrons)$\cdot\pi\cdot 2p^2\cdot\pi^*$	$^1A_1$	$^3A_1$
(Inner electrons)$\cdot\pi^2\cdot 2p\cdot\sigma^*$	$^1B_2$	$^3B_2$
(Inner electrons)$\cdot\pi\cdot 2p^2\cdot\sigma^*$	$^1B_1$	$^3B_1$

We note that all of these states are associated with different configurations of the carbonyl group.   The number of terms which can actually be located spectroscopically for a given molecule is rather severely limited by the selection rules which are discussed in section 7d.   An additional factor is that transitions to states of very high energy occur in the vacuum ultraviolet, where they are likely to be obscured by other effects such as photodissociation or photoionization.   We shall have more to say about the location of terms for formaldehyde in section 7f.

7d SELECTION RULES

A very large number of selection rules, of varying degrees of rigidity, are known for different types of spectroscopic transition. Here we shall describe only the more important ones.

In section 7b we found that the intensity of a transition involving electric dipole radiation is proportional to the square of the transition moment, μ_{mn}, where

$$|\mu_{mn}|^2 = |\mu_{mn}^x|^2 + |\mu_{mn}^y|^2 + |\mu_{mn}^z|^2 \tag{7.38}$$

and

$$\mu_{mn}^x = \int \Psi_m^* \cdot \mu^x \cdot \Psi_n \, d\tau \tag{7.39}$$

is the matrix element of the x component of the dipole moment between the wave functions for the initial and final states of the transition. The dipole moment can be expressed in the form

$$\mu^x = \sum_i e \cdot x_i = e \cdot \bar{x} \tag{7.40}$$

where x_i is the x component of the vector which gives the displacement of electron i from the center of the molecular dipole. Hence the matrix element can be written

$$\mu_{mn}^x = e \int \Psi_m^* \cdot \bar{x} \cdot \Psi_n \, d\tau \tag{7.41}$$

$$= e \cdot X_{mn} \tag{7.42}$$

where X_{mn} is a matrix element of the x coordinate. With the matrix elements in this form we can derive a number of important selection rules by means of symmetry arguments alone.

Consider the integral

$$\int_{-\infty}^{\infty} F(x) \, dx = \int_{-\infty}^{0} F(x) \, dx + \int_{0}^{\infty} F(x) \, dx \tag{7.43}$$

which may be split into two parts as shown.

It is easy to transform this into

$$\int_{-\infty}^{\infty} F(x) \, dx = \int_{0}^{\infty} F(-x) \, dx + \int_{0}^{\infty} F(x) \, dx \tag{7.44}$$

If $F(x)$ is an *odd* function $[F(-x) = -F(x)]$, the integral reduces to zero, while if $F(x)$ is an *even* function $[F(-x) = F(x)]$, the integral is nonzero (Fig. 23).

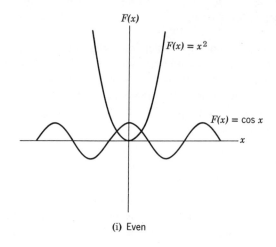

(i) Even

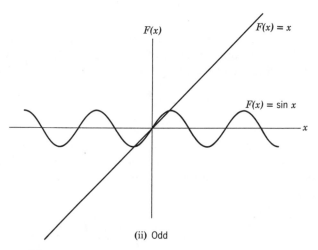

(ii) Odd

Fig. 23   Examples of even and odd functions. (Area above the x axis is positive—area below the x axis is negative—during integration.)

It therefore follows that the matrix element (7.41) is zero if the integrand is an odd function.   Since x itself is an odd function and the product of two odd functions is an even function, this implies that the product of Ψ_m and Ψ_n must also be an odd function of x if the matrix element is to be nonzero.   When we consider μ_{mn}^y

and μ_{mn}^z as well, we see that the product of Ψ_m and Ψ_n must be an odd function with respect to one of the three coordinates x, y, or z if the transition between these states is to be allowed for electric dipole radiation. When only one of the matrix elements is non-zero the transition occurs only with radiation of the appropriate polarization. If two components of the transition moment are non-zero, the radiation is circularly or elliptically polarized.

We can now immediately obtain the *parity selection rule*

$$g \leftrightarrow u \qquad g \nleftrightarrow g \qquad u \nleftrightarrow u \qquad (7.45)$$

Here $\leftrightarrow$ stands for "combines with" and $\nleftrightarrow$ stands for "does not combine with." (This is just another way of saying whether a transition between the pair of states is allowed or not.) This selection rule holds because $\Psi_g \times \Psi_u$ is an odd function and $\Psi_g \times \Psi_g$ and $\Psi_u \times \Psi_u$ are even functions with respect to all three coordinates. The rule applies to all kinds of transition, and one important consequence of this rule is that transitions between states which arise from the same electron configuration are forbidden.

A second rule which concerns symmetry applies to Σ^+ and Σ^- states of diatomic molecules. These states are respectively symmetrical and antisymmetrical with respect to reflection in a (σ_v) plane through the molecular axis. A diatomic molecule can have no component of dipole moment perpendicular to this axis so that in this case we have only to consider the transition moment along the axis, μ_{mn}^x. It is easy to see that the transition moment reduces to zero if the wave functions are of different symmetry types with respect to σ_v, since their product will be positive on one side of the plane and negative on the other, and the two halves will cancel each other out. Hence we obtain the rule

$$\Sigma^+ \leftrightarrow \Sigma^+, \qquad \Sigma^- \leftrightarrow \Sigma^-, \qquad \Sigma^+ \nleftrightarrow \Sigma^- \qquad (7.46)$$

which also applies to 0^+ and 0^- states in Hund's case c.

Rotational energy levels are classified as $(+)$ or $(-)$ according to whether the rotational wave function remains unchanged or changes sign on reflection in the origin (replacement of x, y, z by $-x$, $-y$, $-z$). This is similar to the g, u classification in systems with a center of symmetry, and the same rule holds, that is,

$$+ \leftrightarrow -, \qquad + \nleftrightarrow +, \qquad - \nleftrightarrow - \qquad (7.47)$$

Rotational levels may also possess symmetry with respect to the

interchange of identical nuclei in the molecule. Thus for a homonuclear diatomic molecule the rotational levels are labeled s or a
depending on whether they are symmetric or antisymmetric with
respect to interchange of the nuclei, and in this case the rule is

$$s \leftrightarrow s, \qquad a \leftrightarrow a, \qquad s \nleftrightarrow a \qquad (7.48)$$

In the case of a polyatomic molecule the electronic ground state is
usually totally symmetric, so that a transition to an excited state is
allowed by symmetry if the wave function of this state is odd with
respect to one of the three coordinate axes. This means, in effect,
that the wave function of the excited state should be of the same
symmetry species as x, y, or z. (This is one reason why character
tables usually indicate the representations according to which x, y,
and z transform.) When an electronic transition is expected to be
forbidden on this account it may happen that the transition takes
place with simultaneous excitation of a molecular vibration of the
correct symmetry, and in this case we say that a *vibronic* transition
occurs. Vibronic transitions are particularly important in connection with spectra of complexes and of symmetrical hydrocarbons
such as benzene or anthracene.

With vibrational transitions the condition that the transition
moment should not be zero reduces to the statement that there should
be a change in the dipole moment of the molecule during the vibration. If this condition is fulfilled, the vibration is said to be "infrared active." Again, for transitions from a totally symmetrical
ground vibrational level, this requires that the wave function for the
upper level should be of the same symmetry type as x, y, or z. For a
vibration to be active in the Raman effect the equivalent condition
is that there should be a change of polarizability of the molecule
during a vibration. This is expressed in the statement that the
upper state of the transition should be of the same symmetry species
as one of the components of the *polarizability tensor* (xy, yz, zx,
x^2, y^2, z^2, or some linear combination of these), assuming the lower
state to be totally symmetric. The derivation of the selection rule
for the harmonic oscillator, $\Delta v = \pm 1$ (where the letter v is commonly
used for the vibrational quantum number), was given as an exercise
at the end of Chapter 2. This rule applies generally to molecular
vibrations, except that anharmonicity causes it to be relaxed somewhat, especially when v is large.

Some important selection rules which we shall not attempt to justify are:

(i) $\Delta L = 0, \pm 1$ for atomic states in Russell-Saunders coupling and $\Delta \Lambda = 0, \pm 1$ for diatomic molecules in Hund's case *a* or *b*. (Atomic transitions with $\Delta L = 0$ are usually forbidden by the parity selection rule.)

(ii) For both of these systems we also have $\Delta J = 0, \pm 1$ (but not $0 \leftrightarrow 0$).

(iii) For the molecular Ω (analogous to atomic J) the corresponding rule is $\Delta \Omega = 0, \pm 1$.

(iv) For all cases where S is a good quantum number we have the rule $\Delta S = 0$, that is, transitions between states of different multiplicity are forbidden. This rule applies strictly when only light atoms are involved but breaks down in $j - j$ coupling or Hund's case *c*. Even the physical proximity of a heavy atom can be sufficient to relax the restriction on change of multiplicity.[10]

(v) Electronic transitions are forbidden if they involve the promotion of more than one electron. Allowed transitions are those which involve only a "one-electron jump."

Many other selection rules for atomic and molecular spectra are known and are given in the texts which are listed in Appendix I.[11]

7e MOLECULAR VIBRATIONS

Our aim in this section is to show how the complicated vibrational motion of a polyatomic molecule may be analyzed into a series of relatively simple component vibrational modes, in each of which the molecule behaves, to a good approximation, as a harmonic oscillator. The different modes of vibration are usually known as the "normal modes," or "normal vibrations," and it is possible to define a set of coordinates, known as the "normal coordinates," which express the displacement of the atoms from their equilibrium positions in a

[10] M. Kasha, *J. Chem. Phys.* **20**, 71 (1952). A typical example is the singlet-triplet absorption of 1-chloronaphthalene, which is extremely weak in most solvents but is clearly observed ($\varepsilon \sim 0.01$) in ethyl iodide.

[11] A useful summary, with a discussion of the factors which can lead to violation of selection rules, is contained in the article "Forbidden Transitions" by Garstang, *Atomic and Molecular Processes*, Ed. D. R. Bates, New York: Academic Press, 1962.

particular mode. It is a characteristic of the normal vibrations that the atoms involved all move in phase with one another, each passing through its equilibrium position at the same instant. We shall not consider how the detailed forms of the normal vibrations are determined (for references to this topic see Appendix I), but we will show how it is possible to work out the number of different normal modes of vibration that a molecule may possess, the symmetries of these vibrations, and whether or not they are infrared or Raman active.

The main fact that we shall make use of here is that each normal vibration transforms according to one of the representations of the molecular point group. The reader will already have deduced, from the fact that character tables include reference to x, y, z, R_x, R_y, and R_z, that the same applies to translations and rotations. To illustrate this statement we consider the molecules of H_2O and NH_3.

The point group of H_2O is C_{2v}. The character table is given in Table 7.3, and the symmetry elements are shown in Fig. 24.

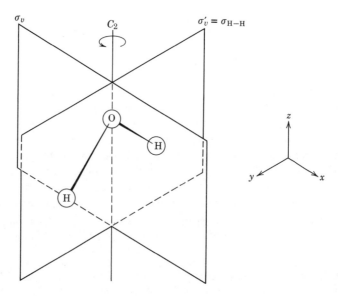

Fig. 24 H_2O, symmetry elements.

Table 7.3

C_{2v}			E	C_2	σ_v	$\sigma_{v'}$
x^2, y^2, z^2	z	A_1	1	1	1	1
xy	R_z	A_2	1	1	-1	-1
xz	R_y, x	B_1	1	-1	1	-1
yz	R_x, y	B_2	1	-1	-1	1

The normal vibrations of H_2O are shown qualitatively in Fig. 25 and the symmetry species of each vibration is indicated. The arrows show the directions in which the atoms move during the vibration. It is easy to verify that the vibration ν_1 is symmetric with respect to E and $\sigma_{v'}$ and antisymmetric with respect to C_2 and σ_v, while ν_2 and ν_3 are symmetric with respect to all four symmetry operations.

The ammonia molecule belongs to point group C_{3v}, for which the character table is shown in Table 7.4.

Table 7.4

C_{3v}			E	$2C_3$	$3\sigma_v$
$x^2 + y^2, z^2$	z	A_1	1	1	1
	R_z	A_2	1	1	-1
$(x^2 - y^2, xy)$ (xz, yz)	(x, y) (R_x, R_y)	E	2	-1	0

The symmetry elements are shown in Fig. 26 and the normal vibrations are shown qualitatively in Fig. 27. In this case we have two pairs of degenerate vibrations, of symmetry type E, which are

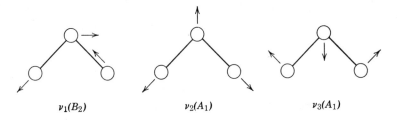

$\nu_1(B_2)$ $\nu_2(A_1)$ $\nu_3(A_1)$

Fig. 25 Normal vibrations of H_2O (not to scale).

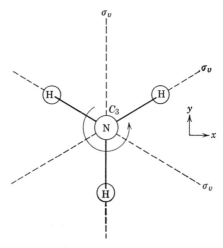

Fig. 26 Ammonia, symmetry elements. (Viewed from above the apex of the triangular pyramid.)

such that the members of a degenerate set are transformed into linear combinations with one another by the action of the symmetry operations. How this occurs may be visualized in terms of the action of the various symmetry operations on x and y, which together form a basis for the E representation in this point group.

The problem now is how to decide the number of normal vibrations of a given symmetry type which are possessed by a molecule. We shall see that this problem, like many others, is essentially one of finding a reducible representation whose irreducible components can be determined with the aid of equation (4.43).

Suppose the molecule in question contains n atoms. Each of these atoms can move in the x, y, or z direction, and any molecular motion can thus be described in terms of $3n$ displacement coordinates. Not all of these $3n$ possible displacements correspond to vibrations since 3 coordinates are required to describe the movement of the whole molecule through space and either 2 or 3 coordinates are required to describe rotations, depending on whether the molecule is linear or nonlinear. Thus the number of degrees of freedom which actually correspond to vibrations is $3n - 5$ for a linear molecule and $3n - 6$ for a nonlinear one, and this is equal to the number

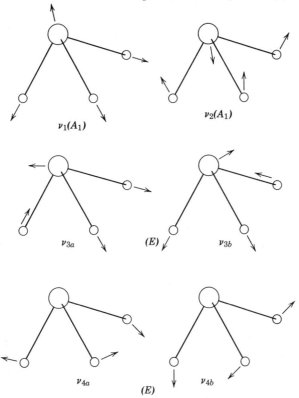

Fig. 27 Normal vibrations of NH_3 (not to scale). For scale diagrams (of ND_3) see Herzberg, *Infra-Red and Raman Spectra*, p. 110. If the atomic displacements in the above diagrams were drawn to scale, the nitrogen atom would appear to be stationary.

of possible modes of vibration. We can form a reducible representation which contains the symmetry species of the molecular rotations, vibration and translations by constructing transformation matrices for the $3n$ cartesian displacement coordinates. These are the matrices which show how the displacement coordinates x, y, and z are transformed by the operation of the symmetry elements of the molecular point group. The problem is simplified by the fact that we do not require the whole matrix of each operation, but only the diagonal elements which make up the character of the operation in the reducible representation.

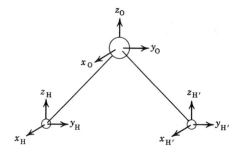

Fig. 28   Displacement coordinates for H_2O.

The displacement coordinates for H_2O are shown in Fig. 28. It is apparent that the only coordinates which will contribute to the character of a symmetry element in the reducible representation are those belonging to an atom whose position is not changed by the symmetry operation. The procedure for finding the characters is therefore similar to that which was used in Chapter 5 for molecular orbitals. This procedure is summarized in Table 7.5.

Table 7.5

Operation $\hat{R}$:	Atoms Left Unmoved:	$\chi(R)_x$		$\chi(R)_y$		$\chi(R)_z$		$\chi(R)$
				Contributions:				
E	H, H', O	3	+	3	+	3	=	9
C_2	O only	-1	+	-1	+	1	=	-1
σ_v	O only	1	+	-1	+	1	=	1
σ_v'	H, H', O	-3	+	3	+	3	=	3

Hence $\Gamma = 3A_1 + A_2 + 2B_1 + 3B_2$ (7.49)

From the character table we find that $A_1 + B_1 + B_2$ correspond to translations and $A_2 + B_1 + B_2$ correspond to rotations, so that we are left with only $2A_1 + B_2$ as vibrations. These are as shown in Fig. 25.

For ammonia the cartesian displacement coordinates are illustrated in Fig. 29.

The determination of the characters now involves a small amount of trigonometry. The process is facilitated if we imagine ourselves

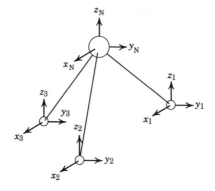

Fig. 29   Cartesian displacement coordinates for NH_3.

looking down on the molecule from a point above the nitrogen atom, as in Fig. 26.   The results are given in Table 7.6.

Table 7.6

Operation:	Unmoved Atom:	$\chi(R)_x$	$\chi(R)_y$	$\chi(R)_z$	$\chi(R)$
		Contributions:			
E	All four	4	4	4	12
σ_v	$\begin{cases} N \\ H_1 \end{cases}$	$\begin{matrix} -1 \\ -1 \end{matrix}$	$\begin{matrix} 1 \\ 1 \end{matrix}$	$\begin{matrix} 1 \\ 1 \end{matrix}$	$\begin{matrix} =1 \\ =1 \end{matrix}\Big\}2$
C_3	N	$-\tfrac{1}{2}$	$-\tfrac{1}{2}$	1	0

Hence $$\Gamma = 3A_1 + A_2 + 4E \qquad (7.50)$$

Subtracting $A_1 + A_2 + 2E$ for translations and rotations leaves $2A_1 + 2E$; the forms of these symmetry vibrations are as shown in Fig. 27.

At ordinary temperatures most molecules are in the ground vibrational state because the spacings between vibrational levels are generally much larger than kT.   The selection rule for the harmonic oscillator (Exercise 2.5) states that the vibrational quantum number may change by only one unit at a time.   Hence if a molecule is irradiated with light of the correct frequency, we expect to observe the transition from $v = 0$ to $v = 1$.   Such transitions, if they are allowed, are responsible for the fundamental infrared absorption bands of molecules. Transitions from levels higher than $v = 0$ give rise to what are usually called "hot bands." It is also sometimes possible to observe "overtones" ($\Delta v = 2, 3, \cdots$) and "combination tones" (two vibrational quantum numbers change simultaneously) but these are usually very weak.

In section 7d we concluded that a transition is allowed if the product $\Psi_m^* \cdot \Psi_n$ is the basis of a representation of the molecular point group which contains x, y, or z. The eigenfunctions of a harmonic oscillator, expressed in terms of the variable $\xi \, (= x\sqrt{\beta})$ of Chapter 2, are such that the ground state eigenfunction has the symmetry of ξ^0, the level with $v = 1$ has the symmetry of ξ, the level with $v = 2$ has the symmetry of ξ^2, and so on. This means that the lowest vibrational eigenfunction is totally symmetric and that the eigenfunction for $v = 1$ has the symmetry of ξ, i.e., it has the symmetry of the appropriate normal coordinate. Hence we can say that a particular vibration will give rise to a fundamental absorption band in the infrared if it belongs to the same symmetry species as x, y, or z. Such a vibration is said to be *infrared active*, and we deduce that all of the normal vibrations of H_2O and NH_3 are active in the infrared.

Vibrational transitions are also commonly observed in the Raman effect, and in this case a fundamental transition is allowed if the vibration is of the same symmetry species as one of the components of the polarizability tensor, x^2, y^2, z^2, xy, yz, zx, or some combination of these. This condition is also fulfilled by the normal vibrations of H_2O and NH_3, so that they are all *Raman active* as well as infrared active.

It often happens that a vibration which is inactive in the infrared can be studied experimentally in the Raman effect, and vice versa (cf. Exercise 7.8). Examples of the application of these methods to other molecules are given in Exercise 7.7.

7f   REPRESENTATIVE SPECTRA OF POLYATOMIC MOLECULES

In this section we shall illustrate the ideas which we have been considering with a discussion of some typical electronic spectra of polyatomic molecules.

The first example which we shall consider is benzene. In section 7c we showed that the ground state configuration $a_{2u}^2 e_{1g}^4$ is of species $^1A_{1g}$ and that the excited configuration $a_{2u}^2 e_{1g}^3 e_{2u}$ gives rise to singlet and triplet states of species B_{1u}, B_{2u}, and E_{1u}. Reference to the character table for D_{6h} shows that transitions to the E_{1u} level are fully allowed, but transitions to $^1B_{1u}$ and $^1B_{2u}$ are forbidden by symmetry. Transitions from $^1A_{1g}$ to the triplet states are all spin-forbidden.

The ultraviolet absorption spectrum of benzene (Fig. 30) contains an intense band ($f = 1.2$) at about 1860 Å which can be assigned to the fully allowed transition $^1E_{1u} \leftarrow {}^1A_{1g}$ (read this as $^1E_{1u}$ *from* $^1A_{1g}$—the upper state is always written first), but there are some other quite striking spectral features in addition to this. The less intense band ($f = 0.002$) at 2600 Å is unusual in that it shows distinct vibrational structure even in solution. This band is attributed to a vibronic transition in which the electronic excitation occurs with simultaneous excitation of a vibration of symmetry E_{1u}, so that the overall transition moment is nonzero. The electronic

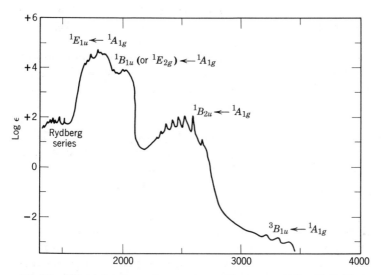

Fig. 30 Ultraviolet absorption spectrum of benzene. (After K. S. Pitzer, *Quantum Chemistry*, Prentice-Hall, New York, 1953.)

transition involved in this case is believed to be $^1B_{2u} \leftarrow {}^1A_{1g}$. The band which forms a shoulder on the main band at 2050 Å ($f = 0.12$) is thought to result from the transition $^1B_{1u} \leftarrow {}^1A_{1g}$, or possibly $^1E_{2g} \leftarrow {}^1A_{1g}$, the symmetry requirement of the transition moment again being met by coupling with an appropriate vibration. Like most organic molecules, benzene also shows absorption at about 1500 Å and below due to Rydberg transitions in which the principle quantum number of a $2p$ electron on a carbon atom is increased by one or more units. These transitions form Rydberg series, similar to the well-known Balmer series for hydrogen, which converge to the first ionization potential of the molecule. Lastly there is a very weak ($f < 7 \times 10^{-12}$) band at 3400 Å which is assigned to a $^3B_{1u} \leftarrow {}^1A_{1g}$ transition. The intensity of this absorption band depends very markedly on the presence of paramagnetic impurities, the difference in intensity between degassed benzene and benzene saturated with oxygen or nitric oxide being by a factor of more than 2000. This observation supports the assignment to a spin-forbidden transition. The upper state is taken to be $^3B_{1u}$ since this is theoretically expected to be the lowest triplet state of benzene.

The triplet states of organic molecules are obviously not very susceptible to study by ordinary spectroscopic techniques. One useful method of observing them is to irradiate a substance in solution or in an organic glass with a flash from a high energy discharge lamp.[12] A small fraction of the molecules which are thus excited to singlet states decays degradatively to nearby triplet states, and it is then possible to observe an absorption spectrum due to allowed triplet-triplet transitions.

We found in section 7c that the first excited configuration of formaldehyde gives rise to $^1A_2$ and $^3A_2$ states. Reference to the C_{2v} character table reveals that transitions from $^1A_1$ to both of these states are forbidden by symmetry, and in addition the transition to $^3A_2$ is obviously spin-forbidden. The observed bands at 2500–3500 Å ($f = 2.4 \times 10^{-4}$) and 3600–4000 Å ($f \simeq 10^{-6}$) are therefore assigned to these transitions. Both bands contain rather long vibrational progressions involving the CO stretching mode, owing to the fact that the CO bond length is greater in the excited state than in the ground state (a π^* orbital is being occupied). Very similar spectra are observed for other carbonyl compounds.

[12] Porter and Windsor, *J. Chem. Phys.*, **21**, 2088 (1953).

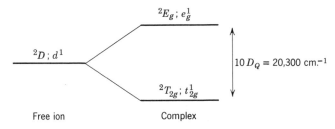

Fig. 31   Correlation diagram for $Ti(H_2O)_6^{3+}$.

We will now consider the spectra of two fairly typical transition metal complexes.

The ion Ti^{3+} has the configuration d^1 and forms octahedral complexes. In aqueous solution the ion is present as violet $Ti(H_2O)_6^{3+}$. The correlation diagram for this complex is given in Fig. 31. It represents the simplest possible application of the crystal field theory which was discussed in Chapter 6.

The ion possesses a crystal field absorption band ($f = 10^{-4}$) at $\bar{\nu} = 20,300$ cm.$^{-1}$, with a shoulder, probably the result of Jahn-Teller distortion, at $\nu = 17,400$ cm.$^{-1}$. We note that all transitions in a complex of O_h symmetry are forbidden by the parity selection rule $g \leftrightarrow g$. Such a transition may become partially allowed as a result either of vibronic coupling or of there being a superimposed field of lower symmetry, e.g., trigonal D_{3h}.

The ion V^{3+} has the d^2 configuration which was considered at some length in Chapter 6. This ion occurs in octahedral coordination in vanadium corundum, which is made by substituting V^{3+} into Al_2O_3. The ground state of the complex is a triplet and the correlation diagram for the triplet states alone is given in Fig. 32.

Intense bands at $\bar{\nu} = 17,400$ cm^{-1}, $25,200$ cm^{-1}, and $34,500$ cm^{-1} are attributed to the transitions $^3T_{2g} \leftarrow {}^3T_{1g}$, $^3T_{1g} \leftarrow {}^3T_{1g}$, and $^3A_{2g} \leftarrow {}^3T_{1g}$ (a, b, and c in Fig. 28). This assignment corresponds to the situation in a strong field, where the upper $^3T_{1g}$ state lies below $^3A_{2g}$. Some weaker bands which have been observed have been assigned to spin-forbidden transitions involving various excited singlet states. For a further example of this type see Exercise 7.9.

It is apparent that the assignment of electronic spectra of polyatomic molecules to transitions between particular pairs of states

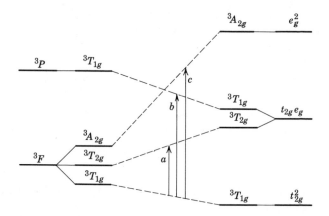

Fig. 32   Correlation diagram for triplet states of V^{3+} in vanadium corundum.

is not always a simple matter, but in most cases a useful guide is provided by the measured f values.   There remains a considerable number of interesting topics which are related to the spectra of polyatomic molecules, for example, charge-transfer spectra, effects of substituents on spectra and nonradiative transitions in fluorescent substances.   For information about these topics the reader is referred to the texts and reviews which are listed in Appendix I.

EXERCISES

7.1   Prove $\mu^x_{mn} = \mu^x_{nm}$.   Hence verify that $B_{mn} = B_{nm}$.

7.2   Consider a 1 cm. cube of a dilute solution which is of concentration $\mathscr{C}$ mols/liter with respect to an absorber.   (The solvent is assumed to be transparent.)   Show that the rate at which quanta are absorbed by the solution is equal to $Q \cdot B_{nm} \cdot \rho(\nu_{mn}) \cdot \mathscr{C} \cdot N/1000$ where Q is the number of quanta per second of frequency ν_{mn} which are incident on the front face of the cube, N is Avogadro's number, and $\rho(\nu_{mn})$ is equal to $Qh\nu_{mn}/c$. Hence derive an expression for the extinction coefficient ε in terms of B_{nm}.

7.3   (*a*) Work out the term symbols for the following states of diatomic molecules.   (*b*) Assign each molecule to one of Hund's coupling cases.

Molecule	S	Λ	Symmetry
LiH	0	0	$+$
O_2	1	0	$g\ -$
NO	$\frac{1}{2}$	1	
N_2^+	$\frac{1}{2}$	0	$g\ +$
ZrO	1	1	
MnBr	3	0	?

7.4 The following electronic states are listed in order of increasing energy. Construct energy level diagrams which show the allowed transitions within each group of states.

   (i) $^2S_\frac{1}{2}$, $^2P_\frac{1}{2}$, $^2P_{1\frac{1}{2}}$, $^2S_\frac{1}{2}$ (Sodium)

   (ii) $^1S_0$, $^3P_0$, $^3P_1$, $^3P_2$, $^3D_1$, $^3D_2$, $^3D_3$, $^1D_2$ (Calcium)

   (iii) $^1\Sigma_g^+$, $^3\Sigma_u^+$, $^3\Pi_g$, $^1\Pi_g$, $^3\Pi_u$, $^1\Pi_u$, $^3\Delta_u$ (Nitrogen)

   (iv) $^3\Sigma_g^-$, $^1\Delta_g$, $^1\Sigma_g^+$, $^3\Sigma_u^+$, $^3\Sigma_u^-$ (Oxygen)

7.5 Work out the symmetry species of the states which arise from the first excited configurations of ethylene and butadiene. Predict which states may be reached by way of allowed transitions from the ground states, and compare the energies of corresponding transitions in the two molecules. (Assume that the singlet-triplet splitting of the excited state is proportionately the same in each case.) It is instructive to carry out this calculation for 6, 8, 10, $\cdots$ carbon atoms in a chain. For the larger molecules the one-dimensional box model is adequate. (Allow a box-length of 1.18Å for each carbon atom, since the chains possess a zig zag configuration.)

7.6 Predict the energies (in terms of β) of allowed transitions to states which arise from the first excited configurations of any of the molecules whose energy levels were determined in the exercises at the end of Chapter 5.

7.7 Work out the symmetry species of the normal vibrations of the following molecules: (i) $BF_3(D_{3h})$; (ii) $C_2H_4(D_{2h})$; (iii) $CoF_6^{3-}(O_h)$. Indicate whether the vibrations are infrared or Raman active in each case. *Answers:*

   (i) A_2'', infrared; A_1', E'', Raman; E', both.

   (ii) B_{1u}, $2B_{2u}$, $2B_{3u}$, infrared; $3A_g$, $2B_{1g}$, B_{2g}, Raman; A_u, inactive.

   (iii) $2T_{1u}$, infrared; A_{1g}, E_g, T_{2g}, Raman; T_{2u}, inactive.

7.8 Compare the g or u symmetry with respect to inversion of x, y, and z with that of the components of the polarizability tensor. Hence prove that in a molecule with a center of symmetry no vibration can be both infrared and Raman active.

7.9   The pale-green octahedral complex of Ni^{II} (d^8 configuration), $Ni(H_2O)_6^{2+}$, has spin-allowed absorption bands at 8500 cm.$^{-1}$, 13,500 cm.$^{-1}$, and 25,300 cm.$^{-1}$. The violet complex $Ni(NH_3)_6^{2+}$ has similar bands at 10,700 cm.$^{-1}$, 17,500 cm.$^{-1}$, and 28,200 cm.$^{-1}$. Assign the bands to particular transitions and compare the values of $10D_q$ for the two complexes.[13]

7.10   The spectroscopic nomenclature of Platt has the advantage over the system based on symmetry classification when it comes to the problem of correlating the ultraviolet spectra of molecules in the series benzene, naphthalene, anthracene, etc.[14]   In this system the energy levels are assumed to be the same as those for an electron confined to a ring, only the number of electrons varying as we go from one member of the series to the next.   (*a*) Show that these wave functions are of the form $\Psi = N \cdot e^{im\phi}$ where ϕ is an angular coordinate and m must be zero or a positive or negative integer.   (*b*) Show that levels with $|m| > 0$ are doubly degenerate.   (*c*) Compare the resulting approximate energy level scheme for benzene with that which was determined in Chapter 5.

[13] Ballhausen (see Appendix I), p. 261.

[14] See Jaffe and Orchin, *Theory and Applications of Ultraviolet Spectroscopy*, New York: Wiley, 1962, p. 302.

Appendix I

Bibliography

....

Chapters 1, 2, and 3. The material which is contained in these chapters is standard quantum mechanics. The present account largely follows Landau and Lifshitz (1), Eyring, Walter, and Kimball (2) and Margenau and Murphy (3). In contrast to the postulatory approach which is followed in Chapter 1 and in the references just given, an interesting inductive approach is used in the book by Eisberg (4). A suitable introductory text to Chapter 1 and to the book as a whole is the book by Coulson (5).

Chapter 4. Molecular symmetry properties and point groups are discussed at about the level of the present account in books by Herzberg (6); Wilson, Decius and Cross (7); Barrow (8); Cotton (9). The first three of these are concerned mainly with the application to molecular vibrations; the last discusses the application of group theory to a variety of chemical problems. The first authors to assume that a knowledge of group theory should be possessed by any serious student of chemistry were Eyring, Walter, and Kimball (2), and our introduction to the subject has many points in common with theirs. A more rigorous and comprehensive treatment of the application of group theory to quantum mechanics is given in the book by Heine (10), where references to the basic mathematical treatises may be found.

Chapter 5. The material covered in this chapter, with many other applications, may also be found in books by Streitwieser (11) and Roberts (12). Some interesting examples are discussed by Cotton

(9).   A slightly different approach is used in the book by Daudel, Lefebvre, and Moser (13).

Chapter 6.   The basic text here is that of Ballhausen (14), while the book by Orgel (15) gives a very good account of the chemical aspects of Ligand Field theory.   Again some very interesting examples are discussed by Cotton (9).   The valence bond approach, which is not dealt with here, is discussed by Pauling in his classic *Nature of the Chemical Bond* (16).   Atomic energy levels are discussed in references (1) and (2) and in the book by Herzberg (17). The fundamental source for the theory of atomic structure is the book by Condon and Shortley (18).

Chapter 7.   Radiation theory is discussed in many of the references already given, e.g., (2), (3), (4), and (8).   The present account largely follows Pitzer (19).   The book by Mitchell and Zemansky (20) is an important reference from the point of view of atomic spectra.   Basic references for the spectra of diatomic molecules are books by Herzberg (21) and Gaydon (22).   For polyatomic molecules we refer mainly to articles by Brand and Williamson (23); Pitts, Wilkinson, and Hammond (24); Mason (25); and to the book by Jaffe and Orchin (26).   The basic text for molecular vibrations is that of Wilson, Delcius, and Cross (7).   A very thorough and detailed account of molecular vibrations is also given by Herzberg (6) and clear introductions are provided by Barrow (8) and Cotton (9), the latter once again with some especially interesting examples.

REFERENCES

1. L. D. Landau and E. M. Lifshitz.   *Quantum Mechanics.   Nonrelativistic Theory*.   Volume 3, *Course in Theoretical Physics*.   Translated from the Russian by J. B. Sykes and J. S. Bell.   Pergamon Press, London, 1959.
2. H. Eyring, J. Walter, and G. E. Kimball.   *Quantum Chemistry*.   John Wiley and Sons, New York, 1944.
3. H. Margenau and G. M. Murphy.   *The Mathematics of Physics and Chemistry*.   2nd Edition.   D. Van Nostrand Co., New Jersey, 1956.
4. R. M. Eisberg.   *Fundamentals of Modern Physics*.   John Wiley and Sons, New York, 1961.
5. C. A. Coulson.   *Valence*.   2nd Edition.   Oxford University Press, London, 1961.

6. G. Herzberg. *Molecular Spectra and Molecular Structure*, Vol. II. *Infra-Red and Raman Spectra of Polyatomic Molecules*. D. Van Nostrand Co., Princeton, New Jersey, 1945.

7. E. B. Wilson, J. C. Decius, and P. C. Cross. *Molecular Vibrations*. McGraw-Hill Book Co., New York, 1955.

8. G. M. Barrow. *Introduction to Molecular Spectroscopy*. McGraw-Hill Book Co., New York, 1962.

9. F. A. Cotton. *Chemical Applications of Group Theory*. Interscience Publishers, a Division of John Wiley, New York, 1963.

10. V. Heine. *Group Theory in Quantum Mechanics*. Pergamon Press, London, 1960.

11. A. Streitwieser. *Molecular Orbital Theory for Organic Chemists*. John Wiley and Sons, New York, 1961.

12. J. D. Roberts. *Notes on Molecular Orbital Calculations*. W. A. Benjamin, New York, 1961.

13. R. Daudel, R. Lefebvre, and C. Moser. *Quantum Chemistry: Methods and Applications*. Interscience Publishers, New York, 1960.

14. C. J. Ballhausen. *Introduction to Ligand Field Theory*. McGraw-Hill Book Co., New York, 1962.

15. L. E. Orgel. *An Introduction to Transition-Metal Chemistry*. Methuen and Co., London, 1960.

16. L. Pauling. *Nature of the Chemical Bond*. 3rd Edition. Cornell University Press, Ithaca, New York, 1960.

17. G. Herzberg. *Atomic Spectra and Atomic Structure*. Dover Publications, New York, 1944.

18. E. U. Condon and G. H. Shortley. *The Theory of Atomic Spectra*. Cambridge University Press, London, 1935.

19. K. S. Pitzer. *Quantum Chemistry*. Prentice Hall, New York, 1953.

20. A. C. G. Mitchell and M. W. Zemansky. *Resonance Radiation and Excited Atoms*. Cambridge University Press, London, 1934.

21. G. Herzberg. *Molecular Spectra and Molecular Structure*, Vol. I, *Spectra of Diatomic Molecules*. 2nd Edition. D. Van Nostrand Co., Princeton, New Jersey, 1950.

22. A. G. Gaydon. *Dissociation Energies*. Chapman and Hall, London, 1953.

23. J. C. D. Brand and D. G. Williamson in *Advances in Physical Organic Chemistry*, Edited by V. Gold. Volume I. Academic Press, London, 1963.

24. J. N. Pitts, F. Wilkinson, and G. S. Hammond, "The 'Vocabulary' of Photochemistry" in *Advances in Photochemistry*. Volume I. Edited by W. A. Noyes, Jr., G. S. Hammond, and J. N. Pitts, Jr. Interscience Publishers, a Division of John Wiley, New York, 1963.

25. S. F. Mason. *Quarterly Reviews of the Chemical Society*. London, **15**, 287 (1961).

26. H. H. Jaffe and M. Orchin. *Theory and Applications of Ultraviolet Spectroscopy*. John Wiley and Sons, New York, 1962.

Appendix II

Selected Character Tables

....

Table A.1

C_1	E
A	1

Table A.2

C_2			E	C_2
x^2, y^2, z^2, xy	R_z, z	A	1	1
xz, yz	R_x, R_y x, y $\}$	B	1	-1

Table A.3

C_3			E	C_3	C_3^2
$x^2 + y^2, z^2$	R_z, z	A	1	1	1
(xz, yz)	(x, y) $\}$	E $\{$	1	ω	ω^2
$(x^2 - y^2, xy)$	(R_x, R_y)		1	ω^2	ω

Note: $\omega = \exp(2\pi i/3)$

Table A.4

C_4			E	C_2	C_4	C_4^3
$x^2 + y^2, z^2$	R_z, z	A	1	1	1	1
$x^2 - y^2, xy$		B	1	1	-1	-1
(xz, yz) $\{$	(x, y) (R_x, R_y) $\}$	E $\{$	1	-1	i	$-i$
			1	-1	$-i$	i

Table A.5

C_5			E	C_5	C_5^2	C_5^3	C_5^4
$x^2 + y^2, z^2$	R_z, z	A	1	1	1	1	1
(xz, yz)	(x, y) (R_x, R_y)	E'	1	ω	ω^2	ω^3	ω^4
			1	ω^4	ω^3	ω^2	ω
$(x^2 - y^2, xy)$		E''	1	ω^2	ω^4	ω	ω^3
			1	ω^3	ω	ω^4	ω^2

Note: $\omega = \exp(2\pi i/5)$

Table A.6

C_6			E	C_6	C_3	C_2	C_3^2	C_6^5
$x^2 + y^2, z^2$	R_z, z	A	1	1	1	1	1	1
		B	1	-1	1	-1	1	-1
(xz, yz)	(x, y) (R_x, R_y)	E'	1	ω	ω^2	ω^3	ω^4	ω^5
			1	ω^5	ω^4	ω^3	ω^2	ω
$(x^2 - y^2, xy)$		E''	1	ω^2	ω^4	1	ω^2	ω^4
			1	ω^4	ω^2	1	ω^4	ω^2

Note: $\omega = \exp(2\pi i/6)$

Table A.7

C_7			E	C_7	C_7^2	C_7^3	C_7^4	C_7^5	C_7^6
$x^2 + y^2, z^2$	z, R_z	A	1	1	1	1	1	1	1
(xz, yz)	(x, y) (R_x, R_y)	E_1	1	ω	ω^2	ω^3	ω^4	ω^5	ω^6
			1	ω^6	ω^5	ω^4	ω^3	ω^2	ω
$(x^2 - y^2, xy)$		E_2	1	ω^2	ω^4	ω^6	ω	ω^3	ω^5
			1	ω^5	ω^3	ω	ω^6	ω^4	ω^2
		E_3	1	ω^3	ω^6	ω^2	ω^5	ω	ω^4
			1	ω^4	ω	ω^5	ω^2	ω^6	ω^3

Note: $\omega = \exp(2\pi i/7)$

Table A.8

C_8			E	C_8	C_4	C_8^3	C_2	C_8^5	C_4^3	C_8^7
$x^2 + y^2, z^2$	z, R_z	A	1	1	1	1	1	1	1	1
		B	1	-1	1	-1	1	-1	1	-1
(xz, yz)	$\left\{\begin{array}{c}(x, y)\\(R_x, R_y)\end{array}\right\}$	E_1	$\begin{cases}1\\1\end{cases}$	$\begin{matrix}\omega\\\omega^7\end{matrix}$	$\begin{matrix}\omega^2\\\omega^6\end{matrix}$	$\begin{matrix}\omega^3\\\omega^5\end{matrix}$	$\begin{matrix}\omega^4\\\omega^4\end{matrix}$	$\begin{matrix}\omega^5\\\omega^3\end{matrix}$	$\begin{matrix}\omega^6\\\omega^2\end{matrix}$	$\begin{matrix}\omega^7\\\omega\end{matrix}$
$(x^2 - y^2, xy)$		E_2	$\begin{cases}1\\1\end{cases}$	$\begin{matrix}\omega^2\\\omega^6\end{matrix}$	$\begin{matrix}\omega^4\\\omega^4\end{matrix}$	$\begin{matrix}\omega^6\\\omega^2\end{matrix}$	$\begin{matrix}1\\1\end{matrix}$	$\begin{matrix}\omega^2\\\omega^6\end{matrix}$	$\begin{matrix}\omega^4\\\omega^4\end{matrix}$	$\begin{matrix}\omega^6\\\omega^2\end{matrix}$
		E_3	$\begin{cases}1\\1\end{cases}$	$\begin{matrix}\omega^5\\\omega^3\end{matrix}$	$\begin{matrix}\omega^2\\\omega^6\end{matrix}$	$\begin{matrix}\omega^7\\\omega\end{matrix}$	$\begin{matrix}\omega^4\\\omega^4\end{matrix}$	$\begin{matrix}\omega\\\omega^7\end{matrix}$	$\begin{matrix}\omega^6\\\omega^2\end{matrix}$	$\begin{matrix}\omega^3\\\omega^5\end{matrix}$

Note: $\omega = \exp(2\pi i/8)$

Table A.9

C_{2v}			E	C_2	σ_v	σ_v'
x^2, y^2, z^2	z	A_1	1	1	1	1
xy	R_z	A_2	1	1	-1	-1
xz	R_y, x	B_1	1	-1	1	-1
yz	R_x, y	B_2	1	-1	-1	1

Table A.10

C_{3v}			E	$2C_3$	$3\sigma_v$
$x^2 + y^2, z^2$	z	A_1	1	1	1
	R_z	A_2	1	1	-1
$(x^2 - y^2, xy)$ (xz, yz)	$\left.\begin{array}{c}(x, y)\\(R_x, R_y)\end{array}\right\}$	E	2	-1	0

Table A.11

C_{4v}			E	C_2	$2C_4$	$2\sigma_v$	$2\sigma_d$
$x^2 + y^2, z^2$	z	A_1	1	1	1	1	1
	R_z	A_2	1	1	1	-1	-1
$x^2 - y^2$		B_1	1	1	-1	1	-1
xy		B_2	1	1	-1	-1	1
(xz, yz)	$\left.\begin{array}{c}(x, y)\\(R_x, R_y)\end{array}\right\}$	E	2	-2	0	0	0

Table A.12

C_{5v}			E	$2C_5$	$2C_5^2$	$5\sigma_v$
$x^2 + y^2, z^2$	z	A_1	1	1	1	1
	R_z	A_2	1	1	1	-1
	$\left.\begin{array}{l}(x, y)\\(R_x, R_y)\end{array}\right\}$	E_1	2	$2\cos\alpha$	$2\cos 2\alpha$	0
$(x^2 - y^2, xy)$		E	2	$2\cos 2\alpha$	$2\cos 4\alpha$	0

Note: $\alpha = 2\pi/5$

Table A.13

C_{6v}			E	C_2	$2C_3$	$2C_6$	$3\sigma_d$	$3\sigma_v$
$x^2 + y^2, z^2$	z	A_1	1	1	1	1	1	1
	R_z	A_2	1	1	1	1	-1	-1
		B_1	1	-1	1	-1	-1	1
		B_2	1	-1	1	-1	1	-1
(xz, yz)	$\left.\begin{array}{l}(x, y)\\(R_x, R_y)\end{array}\right\}$	E_1	2	-2	-1	1	0	0
$(x^2 - y^2, xy)$		E_2	2	2	-1	-1	0	0

Table A.14

C_{1h}			E	σ_h
x^2, y^2, z^2, xy	R_z, x, y	A'	1	1
xz, yz	R_x, R_y, z	A''	1	-1

Table A.15

C_{2h}			E	C_2	σ_h	i
x^2, y^2, z^2, xy	R_z	A_g	1	1	1	1
	z	A_u	1	1	-1	-1
xz, yz	R_x, R_y	B_g	1	-1	-1	1
	x, y	B_u	1	-1	1	-1

Table A.16

$C_{3h} = C_3 \times \sigma_h$ (C_3 plus σ_h)			E	C_3	C_3^2	σ_h	S_3	$S_3' = (S_3)^{-1}$
x^2+y^2, z^2	R_z	A'	1	1	1	1	1	1
	z	A''	1	1	1	-1	-1	-1
(x^2-y^2, xy)	(x, y)	E' $\begin{cases} \\ \\ \end{cases}$	1	ω	ω^2	1	ω	ω^2
			1	ω^2	ω	1	ω^2	ω
(xz, yz)	(R_x, R_y)	E'' $\begin{cases} \\ \\ \end{cases}$	1	ω	ω^2	-1	$-\omega$	$-\omega^2$
			1	ω^2	ω	-1	$-\omega^2$	$-\omega$

Note: $\omega = \exp(2\pi i/3)$

Table A.17

$C_{4h} = C_4 \times i$			E	C_2	C_4	C_4^3	i	σ_h	S_4	S_4^3
x^2+y^2, z^2	R_z	A_g	1	1	1	1	1	1	1	1
x^2-y^2, xy		B_g	1	1	-1	-1	1	1	-1	-1
(xz, yz)	(R_x, R_y)	E_g $\begin{cases} \\ \\ \end{cases}$	1	-1	i	$-i$	1	-1	i	$-i$
			1	-1	$-i$	i	1	-1	$-i$	i
	z	A_u	1	1	1	1	-1	-1	-1	-1
		B_u	1	1	-1	-1	-1	-1	1	1
	(x, y)	E_u $\begin{cases} \\ \\ \end{cases}$	1	-1	i	$-i$	-1	1	$-i$	i
			1	-1	$-i$	i	-1	1	i	$-i$

A.18 $C_{5h} = C_5 \times \sigma_h$

A.19 $C_{6h} = C_6 \times i$

Table A.20

S_2			E	i
$x^2, y^2, z^2, xy, xz, yz$	R_x, R_y, R_z	A_g	1	1
	x, y, z	A_u	1	-1

Table A.21

S_4			E	C_2	S_4	S_4^3
x^2+y^2, z^2	R_z	A	1	1	1	1
	z	B	1	1	-1	-1
(xz, yz)	(x, y)	E $\begin{cases} \\ \\ \end{cases}$	1	-1	i	$-i$
(x^2-y^2, xy)	(R_x, R_y)		1	-1	$-i$	i

A.22 $S_6 = C_3 \times i$

Table A.23

D_2			E	C_2^x	C_2^y	C_2^z
x^2, y^2, z^2		A_1	1	1	1	1
xy	R_z, z	B_1	1	1	-1	-1
xz	R_y, y	B_2	1	-1	1	-1
yz	R_x, x	B_3	1	-1	-1	1

Table A.24

D_3			E	$2C_3$	$3C_2'$
$x^2 + y^2, z^2$		A_1	1	1	1
	R_z, z	A_2	1	1	-1
(xz, yz) $(x^2 - y^2, xy)$	(x, y) (R_x, R_y)	E	2	-1	0

Table A.25

D_4			E	C_2	$2C_4$	$2C_2'$	$2C_2''$
$x^2 + y^2, z^2$		A_1	1	1	1	1	1
	R_z, z	A_2	1	1	1	-1	-1
		B_1	1	1	-1	1	-1
		B_2	1	1	-1	-1	1
(xz, yz) $(x^2 - y^2, xy)$	(x, y) (R_x, R_y)	E	2	-2	0	0	0

Table A.26

D_5			E	$2C_5$	$2C_5^2$	$5C_2'$
$x^2 + y^2, z^2$		A_1	1	1	1	1
	R_z, z	A_2	1	1	1	-1
(xz, yz)	(x, y) (R_x, R_y)	E_1	2	$2\cos\alpha$	$2\cos 2\alpha$	0
$(x^2 - y^2, xy)$		E_2	2	$2\cos 2\alpha$	$2\cos 4\alpha$	0

Note: $\alpha = 2\pi/5$

Table A.27

D_6			E	C_2	$2C_3$	$2C_6$	$3C_2'$	$3C_2''$
$x^2 + y^2, z^2$		A_1	1	1	1	1	1	1
	R_z, z	A_2	1	1	1	1	-1	-1
		B_1	1	-1	1	-1	1	-1
		B_2	1	-1	1	-1	-1	1
(xz, yz)	$\left.\begin{array}{c}(x, y)\\(R_x, R_y)\end{array}\right\}$	E_1	2	-2	-1	1	0	0
$(x^2 - y^2, xy)$		E_2	2	2	-1	-1	0	0

Table A.28

D_{2d}			E	C_2	$2S_4$	$2C_2'$	$2\sigma_d$
$x^2 + y^2, z^2$		A_1	1	1	1	1	1
	R_z	A_2	1	1	1	-1	-1
$x^2 - y^2$		B_1	1	1	-1	1	-1
xy	z	B_2	1	1	-1	-1	1
(xz, yz)	$\left.\begin{array}{c}(x, y)\\(R_x, R_y)\end{array}\right\}$	E	2	-2	0	0	0

A.29 $D_{3d} = D_3 \times i$

A.30 $D_{2h} = D_2 \times i$

Table A.31

$D_{3h} = D_3 \times \sigma_h$			E	σ_h	$2C_3$	$2S_3$	$3C_2$	$3\sigma_v$
$x^2 + y^2, z^2$		A_1'	1	1	1	1	1	1
	R_z	A_2'	1	1	1	1	-1	-1
		A_1''	1	-1	1	-1	1	-1
	z	A_2''	1	-1	1	-1	-1	1
$(x^2 - y^2, xy)$	(x, y)	E'	2	2	-1	-1	0	0
(xz, yz)	(R_x, R_y)	E''	2	-2	-1	1	0	0

A.32 $D_{4h} = D_4 \times i$

A.33 $D_{5h} = D_5 \times \sigma_h$

A.34 $D_{6h} = D_6 \times i$

Table A.35

T				E	$3C_2$	$4C_3$	$4C_3'$
$x^2 + y^2 + z^2$			A	1	1	1	1
$(2z^2 - x^2 - y^2,$			E $\{$	1	1	ω	ω^2
$x^2 - y^2)$				1	1	ω^2	ω
(xy, yz, zx)	$\{$	(x, y, z) (R_x, R_y, R_z) $\}$ T		3	-1	0	0

Note: $\omega = \exp(2\pi i/3)$

Table A.36

T_d				E	$8C_3$	$3C_2$	$6\sigma_d$	$6S_4$
$x^2 + y^2 + z^2$			A_1	1	1	1	1	1
			A_2	1	1	1	-1	-1
$(2z^2 - x^2 - y^2, x^2 - y^2)$			E	2	-1	2	0	0
		(R_x, R_y, R_z)	T_1	3	0	-1	1	-1
(xy, yz, zx)		(x, y, z)	T_2	3	0	-1	-1	1

Table A.37

O				E	$8C_3$	$3C_2$	$6C_2'$	$6C_4$
$x^2 + y^2 + z^2$			A_1	1	1	1	1	1
			A_2	1	1	1	-1	-1
$(x^2 - y^2, 2z^2 - x^2 - y^2)$			E	2	-1	2	0	0
		(R_x, R_y, R_z)	T_1	3	0	-1	-1	1
(xy, yz, zx)		(x, y, z)	T_2	3	0	-1	1	-1

A.38 $O_h = O \times i$

Table A.39

$C_{\infty v}$				E	$2C_\alpha$	σ_v
$x^2 + y^2, z^2$			$A_1 = \Sigma^+$	1	1	1
	z		$A_2 = \Sigma^-$	1	1	-1
(xz, yz)	$\{$	(x, y) (R_x, R_y)	$E_1 = \Pi$	2	$2\cos\alpha$	0
$(x^2 - y^2, xy)$			$E_2 = \Delta$	2	$2\cos 2\alpha$	0
			$E_3 = \Phi$	2	$2\cos 3\alpha$	0
			$\cdot \quad \cdot \quad \cdot$	$\cdot$	$\cdot \quad \cdot \quad \cdot \quad \cdot$	$\cdot$

Table A.40

$D_{\infty h} = C_{\infty v} \times i$			E	$2C_\alpha$	$C_2' = \sigma_v$	i	$2iC_\alpha$	iC_2'
x^2+y^2, z^2		$A_{1g} = \Sigma_g^+$	1	1	1	1	1	1
		$A_{1u} = \Sigma_u^+$	1	1	1	-1	-1	-1
		$A_{2g} = \Sigma_g^-$	1	1	-1	1	1	-1
	z	$A_{2u} = \Sigma_u^-$	1	1	-1	-1	-1	1
(xz, yz)	(R_x, R_y)	$E_{1g} = \Pi_g$	2	$2\cos\alpha$	0	2	$2\cos\alpha$	0
	(x, y)	$E_{1u} = \Pi_u$	2	$2\cos\alpha$	0	-2	$-2\cos\alpha$	0
(x^2-y^2, xy)		$E_{2g} = \Delta_g$	2	$2\cos 2\alpha$	0	2	$2\cos 2\alpha$	0
		$E_{2u} = \Delta_u$	2	$2\cos 2\alpha$	0	-2	$-2\cos 2\alpha$	0
		. . . . .	. . . .					

Index